Introduction to Heat Transfer

Introduction to Heat Transfer

Edited by
Nathaniel Harris

Introduction to Heat Transfer
Edited by Nathaniel Harris
ISBN: 978-1-63549-142-5 (Hardback)

© 2017 Larsen & Keller

Published by Larsen and Keller Education,
5 Penn Plaza,
19th Floor,
New York, NY 10001, USA

Cataloging-in-Publication Data

Introduction to heat transfer / edited by Nathaniel Harris.
p. cm.
Includes bibliographical references and index.
ISBN 978-1-63549-142-5
1. Heat--Transmission. 2. Energy transfer. 3. Thermodynamics.
I. Harris, Nathaniel.
QC320 .I58 2017
621.402 2--dc23

This book contains information obtained from authentic and highly regarded sources. All chapters are published with permission under the Creative Commons Attribution Share Alike License or equivalent. A wide variety of references are listed. Permissions and sources are indicated; for detailed attributions, please refer to the permissions page. Reasonable efforts have been made to publish reliable data and information, but the authors, editors and publisher cannot assume any responsibility for the vailidity of all materials or the consequences of their use.

Trademark Notice: All trademarks used herein are the property of their respective owners. The use of any trademark in this text does not vest in the author or publisher any trademark ownership rights in such trademarks, nor does the use of such trademarks imply any affiliation with or endorsement of this book by such owners.

The publisher's policy is to use permanent paper from mills that operate a sustainable forestry policy. Furthermore, the publisher ensures that the text paper and cover boards used have met acceptable environmental accreditation standards.

Printed and bound in the United States of America.

For more information regarding Larsen and Keller Education and its products, please visit the publisher's website www.larsen-keller.com

Table of Contents

Preface

This book unfolds the innovative aspects of heat transfer which will be crucial for the holistic understanding of the subject of heat transfer. It is designed in such a way that it provides a detailed explanation of the various concepts and applications of this subject matter. Heat transfer refers to the process when two or more physical systems exchange thermal energy. The aim of this textbook is to make the complex subject of heat transfer easy to comprehend and understand. The topics included in this text are of utmost significance and bound to provide incredible insights to readers. The various applications along with technological progress that have future implications are glanced at in it. Those in search of information to further their knowledge will be greatly assisted by this textbook. It will greatly assist students in the fields of chemical engineering, kinetics and computational modeling.

A detailed account of the significant topics covered in this book is provided below:

Chapter 1- The transference of heat between two disparate thermodynamic systems may be termed as heat transfer. The three fundamental modes of heat transfer are conduction, convection and radiation. This chapter will provide an integrated understanding of heat transfer.

Chapter 2- The energy that is passed between a system and its surroundings is termed as heat; heat usually transfers from a hotter to a colder body. The following chapter will not only provide an overview, it will also examine into the variegated topics related to it.

Chapter 3- The transferring of heat by microscopic collisions within a body is known as thermal conduction. Conduction as a process takes place in every phase of matter, solids, liquids, gases and plasma. The alternative modes of heat transfer explained in this section are convective heat transfer and radiation. The topics discussed in the chapter are of great importance to broaden the existing knowledge on heat transfer.

Chapter 4- The property of a material to conduct heat is thermal conductivity whereas the change of shape and area of matter in response to temperature change through the phenomenon of heat conductivity is thermal expansion. Thermal equilibrium and its relation to heat transfer is also elucidated in the following chapter.

Chapter 5- This chapter aims to stretch the boundaries of physics by clarifying and developing the theoretical and conceptual framework of physics. The fundamental theories such as the first law of thermodynamics, second law of thermodynamics, zeroth law of thermodynamics and NTU method are discussed in this chapter, providing a better understanding of heat transfer.

Chapter 6- Tools and techniques are an important component of any field of study. The following chapter elucidates the various tools and techniques that are related to heat transfer. It serves as a source to understand the major categories of heat transfer, such as heat exchanger, heat spreader, heat sink and loop heat pipe. The aspects elucidated are of vital importance, and provide a better understanding of heat transfer.

It gives me an immense pleasure to thank our entire team for their efforts. Finally in the end, I would like to thank my family and colleagues who have been a great source of inspiration and support.

Editor

1

Introduction to Heat Transfer

The transference of heat between two disparate thermodynamic systems may be termed as heat transfer. The three fundamental modes of heat transfer are conduction, convection and radiation. This chapter will provide an integrated understanding of heat transfer.

Heat transfer is the exchange of thermal energy between physical systems. The rate of heat transfer is dependent on the temperatures of the systems and the properties of the intervening medium through which the heat is transferred. The three fundamental modes of heat transfer are *conduction*, *convection* and *radiation*. Heat transfer, the flow of energy in the form of heat, is a process by which a system changes its internal energy, hence is of vital use in applications of the First Law of Thermodynamics. Conduction is also known as diffusion, not to be confused with diffusion related to the mixing of constituents of a fluid.

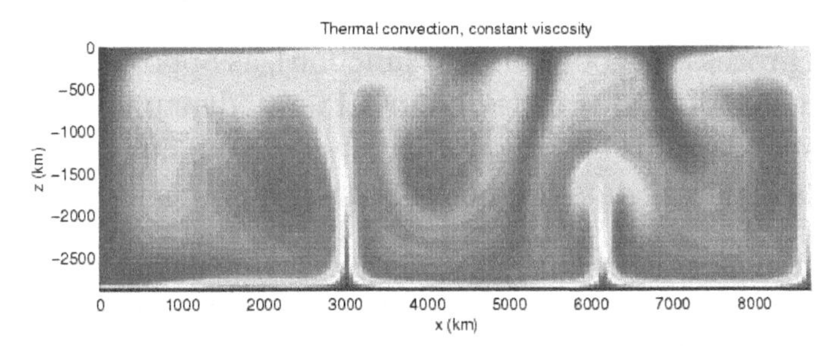

This figure shows a calculation for thermal convection in the Earth's mantle. Colors closer to red are hot areas and colors closer to blue are cold areas. A hot, less-dense lower boundary layer sends plumes of hot material upwards, and likewise, cold material from the top moves downwards.

The direction of heat transfer is from a region of high temperature to another region of lower temperature, and is governed by the Second Law of Thermodynamics. Heat transfer changes the internal energy of the systems from which and to which the energy is transferred. Heat transfer will occur in a direction that increases the entropy of the collection of systems.

Thermal equilibrium is reached when all involved bodies and the surroundings reach the same temperature. Thermal expansion is the tendency of matter to change in volume in response to a change in temperature.

Overview

Heat is defined in physics as the transfer of thermal energy across a well-defined boundary around a thermodynamic system. The thermodynamic free energy is the amount of work that a thermodynamic system can perform. Enthalpy is a thermodynamic potential, designated by the letter "H", that is the sum of the internal energy of the system (U) plus the product of pressure (P) and volume (V). Joule is a unit to quantify energy, work, or the amount of heat.

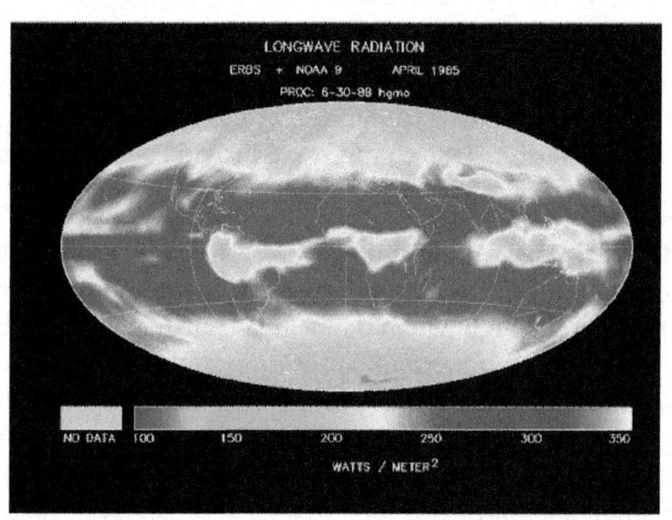

Earth's longwave thermal radiation intensity, from clouds, atmosphere and surface.

Heat transfer is a process function (or path function), as opposed to functions of state; therefore, the amount of heat transferred in a thermodynamic process that changes the state of a system depends on how that process occurs, not only the net difference between the initial and final states of the process.

Thermodynamic and mechanical heat transfer is calculated with the heat transfer coefficient, the proportionality between the heat flux and the thermodynamic driving force for the flow of heat. Heat flux is a quantitative, vectorial representation of heat-flow through a surface.

In engineering contexts, the term *heat* is taken as synonymous to thermal energy. This usage has its origin in the historical interpretation of heat as a fluid (*caloric*) that can be transferred by various causes, and that is also common in the language of laymen and everyday life.

The transport equations for thermal energy (Fourier's law), mechanical momentum (Newton's law for fluids), and mass transfer (Fick's laws of diffusion) are similar, and analogies among these three transport processes have been developed to facilitate prediction of conversion from any one to the others.

Thermal engineering concerns the generation, use, conversion, and exchange of heat transfer. As such, heat transfer is involved in almost every sector of the economy. Heat transfer is classified into various mechanisms, such as thermal conduction, thermal convection, thermal radiation, and transfer of energy by phase changes.

Mechanisms

The fundamental modes of heat transfer are:

Advection

Advection is the transport mechanism of a fluid from one location to another, and is dependent on motion and momentum of that fluid.

Conduction or diffusion

The transfer of energy between objects that are in physical contact. Thermal conductivity is the property of a material to conduct heat and evaluated primarily in terms of Fourier's Law for heat conduction.

Convection

The transfer of energy between an object and its environment, due to fluid motion. The average temperature, is a reference for evaluating properties related to convective heat transfer.

Radiation

The transfer of energy by the emission of electromagnetic radiation.

Advection

By transferring matter, energy—including thermal energy—is moved by the physical transfer of a hot or cold object from one place to another. This can be as simple as placing hot water in a bottle and heating a bed, or the movement of an iceberg in changing ocean currents. A practical example is thermal hydraulics.This can be described by the formula:

$$Q = v\rho c_p \Delta T$$

where Q is heat flux (W/m^2), ρ is density (kg/m^3), c_p is heat capacity at constant pressure (J/kg·K), ΔT is the change in temperature (K), v is velocity (m/s).

Conduction

On a microscopic scale, heat conduction occurs as hot, rapidly moving or vibrating atoms and molecules interact with neighboring atoms and molecules, transferring some of their energy (heat) to these neighboring particles. In other words, heat is transferred by conduction when adjacent atoms vibrate against one another, or as electrons move from one atom to another. Conduction is the most significant means of heat transfer within a solid or between solid objects in thermal contact. Fluids—especially gases—are less conductive. Thermal contact conductance is the study of heat conduction between solid bodies in contact. The process of heat transfer from one place to another place without the movement of particles is called conduction Example:Heat transfer through Metal rods. *Steady state conduction* is a form of conduction that happens when the temperature difference driving the conduction is constant, so that after an equilibration time, the spatial distribution of temperatures in the conducting object does not change any further. In steady state conduction, the amount of heat entering a section is equal to amount of heat coming out.

Transient conduction occurs when the temperature within an ob-ject changes as a function of time. Analysis of transient systems is more complex and often calls for the application of approximation theories or numerical analysis by computer.

Convection

The flow of fluid may be forced by external processes, or sometimes (in gravitational fields) by buoyancy forces caused when thermal energy expands the fluid (for example in a fire plume), thus influencing its own transfer. The latter process is often called "natural convection". All convective processes also move heat partly by diffusion, as well. Another form of convection is forced convection. In this case the fluid is forced to flow by use of a pump, fan or other mechanical means.

Convective heat transfer, or convection, is the transfer of heat from one place to another by the movement of fluids, a process that is essentially the transfer of heat via mass transfer. Bulk motion of fluid enhances heat transfer in many physical situations, such as (for example) between a solid surface and the fluid. Convection is usually the dominant form of heat transfer in liquids and gases. Although sometimes discussed as a third method of heat transfer, convection is usually used to describe the combined effects of heat conduction within the fluid (diffusion) and heat transference by bulk fluid flow streaming. The process of transport by fluid streaming is known as advection, but pure advection is a term that is generally associated only with mass transport in fluids, such as advection of pebbles in a river. In the case of heat transfer in fluids, where transport by advection in a fluid is always also accompanied by transport via heat diffusion (also known as heat conduction) the process of heat convection is understood to refer to the sum of heat transport by advection and diffusion/conduction.

Free, or natural, convection occurs when bulk fluid motions (streams and currents) are caused by buoyancy forces that result from density variations due to variations of temperature in the fluid. *Forced* convection is a term used when the streams and currents in the fluid are induced by external means—such as fans, stirrers, and pumps—creating an artificially induced convection current.

Convection-cooling

Convective cooling is sometimes described as Newton's law of cooling:

The rate of heat loss of a body is proportional to the temperature difference between the body and its surroundings.

However, by definition, the validity of Newton's law of cooling requires that the rate of heat loss from convection be a linear function of ("proportional to") the temperature difference that drives heat transfer, and in convective cooling this is sometimes not the case. In general, convection is not linearly dependent on temperature gradients, and in some cases is strongly nonlinear. In these cases, Newton's law does not apply.

Convection vs. Conduction

In a body of fluid that is heated from underneath its container, conduction and convection can be considered to compete for dominance. If heat conduction is too great, fluid moving down by convection is heated by conduction so fast that its downward movement will be stopped due to its buoyancy, while fluid moving up by convection is cooled by conduction so fast that its driving buoyancy will diminish. On the other hand, if heat conduction is very low, a large temperature gradient may be formed and convection might be very strong.

The Rayleigh number (Ra) is a measure determining the relative strength of conduction and convection.

$$Ra = \frac{g\Delta\rho L^3}{\mu\alpha} = \frac{g\beta\Delta T L^3}{\nu\alpha}$$

where

- g is acceleration due to gravity,
- ρ is the density with Äρ being the density difference between the lower and upper ends,
- μ is the dynamic viscosity,
- α is the Thermal diffusivity,
- β is the volume thermal expansivity (sometimes denoted α elsewhere),
- T is the temperature,
- ν is the kinematic viscosity, and
- L is characteristic length.

The Rayleigh number can be understood as the ratio between the rate of heat transfer by convection to the rate of heat transfer by conduction; or, equivalently, the ratio between the corresponding timescales (i.e. conduction timescale divided by convection timescale), up to a numerical factor. This can be seen as follows, where all calculations are up to numerical factors depending on the geometry of the system.

The buoyancy force driving the convection is roughly $g\Delta\rho L^3$, so the corresponding pressure is roughly $g\ddot{A}\rho L$. In steady state, this is canceled by the shear stress due to viscosity, and therefore roughly equals $\mu V / L = \mu / T_{conv}$, where V is the typical fluid velocity due to convection and T_{conv} the order of its timescale.The conduction timescale, on the other hand, is of the order of $T_{cond} = L^2 / \alpha$.

Convection occurs when the Rayleigh number is above 1,000–2,000.

Radiation

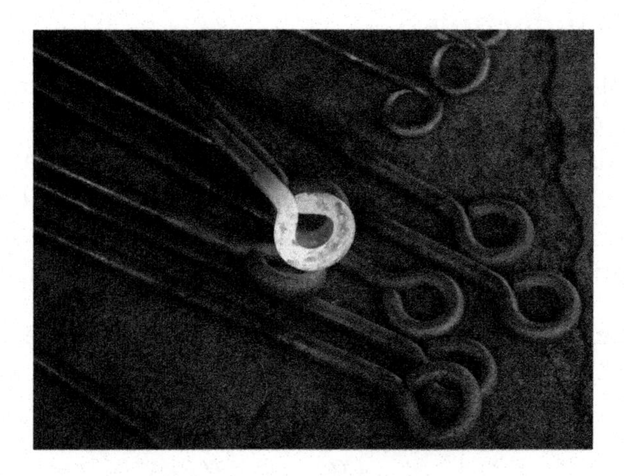

Red-hot iron object, transferring heat to the surrounding environment primarily through thermal radiation

Thermal radiation occurs through a vacuum or any transparent medium (solid or fluid). It is the transfer of energy by means of photons in electromagnetic waves governed by the same laws. Earth's radiation balance depends on the incoming and the outgoing thermal radiation, Earth's energy budget. Anthropogenic perturbations in the climate system are responsible for a positive radiative forcing which reduces the net longwave radiation loss to space.

Thermal radiation is energy emitted by matter as electromagnetic waves, due to the pool of thermal energy in all matter with a temperature above absolute zero. Thermal radiation propagates without the presence of matter through the vacuum of space.

Thermal radiation is a direct result of the random movements of atoms and molecules in matter. Since these atoms and molecules are composed of charged particles (protons and electrons), their movement results in the emission of electromagnetic radiation, which carries energy away from the surface.

The Stefan-Boltzmann equation, which describes the rate of transfer of radiant energy, is as follows for an object in a vacuum :

$$Q = \epsilon \sigma T^4$$

For radiative transfer between two objects, the equation is as follows:

$$Q = \epsilon\sigma(T_a^4 - T_b^4)$$

where Q is the heat flux, ε is the emissivity (unity for a black body), σ is the Stefan-Boltzmann constant, and T is the absolute temperature (in Kelvin or Rankine). Radiation is typically only important for very hot objects, or for objects with a large temperature difference.

Radiation from the sun, or solar radiation, can be harvested for heat and power. Unlike conductive and convective forms of heat transfer, thermal radiation can be concentrated in a small spot by using reflecting mirrors, which is exploited in concentrating solar power generation. For example, the sunlight reflected from mirrors heats the PS10 solar power tower and during the day it can heat water to 285 °C (545 °F).

Phase Transition

Lightning is a highly visible form of energy transfer and is an example of plasma present at Earth's surface. Typically, lightning discharges 30,000 amperes at up to 100 million volts, and emits light, radio waves, X-rays and even gamma rays. Plasma temperatures in lightning can approach 28,000 Kelvin (27,726.85 °C) (49,940.33 °F) and electron densities may exceed 1024 m−3.

Phase transition or phase change, takes place in a thermodynamic system from one phase or state of matter to another one by heat transfer. Phase change examples are the melting of ice or the boiling of water. The Mason equation explains the growth of a water droplet based on the effects of heat transport on evaporation and condensation.

Types of phase transition occurring in the four fundamental states of matter, include:

- Solid - Deposition, freezing and solid to solid transformation.
- Gas - Boiling/evaporation, recombination/deionization, and sublimation.
- Liquid - Condensation and melting/fusion.
- Plasma - Ionization.

Boiling

The boiling point of a substance is the temperature at which the vapor pressure of the liquid equals the pressure surrounding the liquid and the liquid evaporates resulting in an abrupt change in vapor volume.

Nucleate boiling of water.

Saturation temperature means boiling point. The saturation temperature is the temperature for a corresponding saturation pressure at which a liquid boils into its vapor phase. The liquid can be said to be saturated with thermal energy. Any addition of thermal energy results in a phase transition.

At standard atmospheric pressure and low temperatures, no boiling occurs and the heat transfer rate is controlled by the usual single-phase mechanisms. As the surface temperature is increased, local boiling occurs and vapor bubbles nucleate, grow into the surrounding cooler fluid, and collapse. This is *sub-cooled nucleate boiling*, and is a very efficient heat transfer mechanism. At high bubble generation rates, the bubbles begin to interfere and the heat flux no longer increases rapidly with surface temperature (this is the departure from nucleate boiling, or DNB).

At similar standard atmospheric pressure and high temperatures, the hydrodynamically-quieter regime of film boiling is reached. Heat fluxes across the stable vapor layers are low, but rise slowly with temperature. Any contact between fluid and the surface that may be seen probably leads to the extremely rapid nucleation of a fresh vapor layer ("spontaneous nucleation"). At higher temperatures still, a maximum in the heat flux is reached (the critical heat flux, or CHF).

The Leidenfrost Effect demonstrates how nucleate boiling slows heat transfer due to gas bubbles on the heater's surface. As mentioned, gas-phase thermal conductivity is much lower than liquid-phase thermal conductivity, so the outcome is a kind of "gas thermal barrier".

Condensation

Condensation occurs when a vapor is cooled and changes its phase to a liquid. During condensation, the latent heat of vaporization must be released. The amount of the heat is the same as that absorbed during vaporization at the same fluid pressure.

There are several types of condensation:

- Homogeneous condensation, as during a formation of fog.
- Condensation in direct contact with subcooled liquid.
- Condensation on direct contact with a cooling wall of a heat exchanger: This is the most common mode used in industry:
    - Filmwise condensation is when a liquid film is formed on the subcooled surface, and usually occurs when the liquid wets the surface.
    - Dropwise condensation is when liquid drops are formed on the subcooled surface, and usually occurs when the liquid does not wet the surface.

Dropwise condensation is difficult to sustain reliably; therefore, industrial equipment is normally designed to operate in filmwise condensation mode.

Melting

Ice melting

Melting is a physical process that results in the phase transition of a substance from a solid to a liquid. The internal energy of a substance is increased, typically by the application of heat or pressure, resulting in a rise of its temperature to the melting point, at which the ordering of ionic or molecular entities in the solid breaks down to a less ordered state and the solid liquefies. An object that has melted completely is molten. Substances in the molten state generally have reduced viscosity with elevated temperature; an exception to this maxim is the element sulfur, whose viscosity increases to a point due to polymerization and then decreases with higher temperatures in its molten state.

Modeling Approaches

Heat transfer can be modeled in the following ways.

Climate Models

Climate models study the radiant heat transfer by using quantitative methods to simulate the interactions of the atmosphere, oceans, land surface, and ice.

Heat Equation

The heat equation is an important partial differential equation that describes the distribution of heat (or variation in temperature) in a given region over time. In some cases, exact solutions of the equation are available; in other cases the equation must be solved numerically using computational methods.

Lumped System Analysis

Lumped system analysis often reduces the complexity of the equations to one first-order linear differential equation, in which case heating and cooling are described by a simple exponential solution, often referred to as Newton's law of cooling.

System analysis by the lumped capacitance model is a common approximation in transient conduction that may be used whenever heat conduction within an object is much faster than heat conduction across the boundary of the object. This is a method of approximation that reduces one aspect of the transient conduction system—that within the object—to an equivalent steady state system. That is, the method assumes that the temperature within the object is completely uniform, although its value may be changing in time.

In this method, the ratio of the conductive heat resistance within the object to the convective heat transfer resistance across the object's boundary, known as the *Biot number*, is calculated. For small Biot numbers, the approximation of *spatially uniform temperature within the object* can be used: it can be presumed that heat transferred into the object has time to uniformly distribute itself, due to the lower resistance to doing so, as compared with the resistance to heat entering the object.

Engineering

Heat transfer has broad application to the functioning of numerous devices and systems.

Heat-transfer principles may be used to preserve, increase, or decrease temperature in a wide variety of circumstances.Heat transfer methods are used in numerous disciplines, such as automotive engineering, thermal management of electronic devices and systems, climate control, insulation, materials processing, and power station engineering.

Heat exposure as part of a fire test for firestop products

Insulation, Radiance and Resistance

Thermal insulators are materials specifically designed to reduce the flow of heat by limiting conduction, convection, or both. Thermal resistance is a heat property and the measurement by which an object or material resists to heat flow (heat per time unit or thermal resistance) to temperature difference.

Radiance or spectral radiance are measures of the quantity of radiation that passes through or is emitted. Radiant barriers are materials that reflect radiation, and therefore reduce the flow of heat from radiation sources. Good insulators are not necessarily good radiant barriers, and vice versa. Metal, for instance, is an excellent reflector and a poor insulator.

The effectiveness of a radiant barrier is indicated by its reflectivity, which is the fraction of radiation reflected. A material with a high reflectivity (at a given wavelength) has a low emissivity (at that same wavelength), and vice versa. At any specific wavelength, reflectivity = 1 - emissivity. An ideal radiant barrier would have a reflectivity of 1, and would therefore reflect 100 percent of incoming radiation. Vacuum flasks, or Dewars, are silvered to approach this ideal. In the vacuum of space, satellites use multi-layer insulation, which consists of many layers of aluminized (shiny) Mylar to greatly reduce radiation heat transfer and control satellite temperature.

Devices

- Heat engine is a system that performs the conversion of heat or thermal energy to mechanical energy which can then be used to do mechanical work.

- Thermocouple is a temperature-measuring device and widely used type of temperature sensor for measurement and control, and can also be used to convert heat into electric power.
- Thermoelectric cooler is a solid state electronic device that pumps (transfers) heat from one side of the device to the other when electric current is passed through it. It is based on the Peltier effect.
- Thermal diode or thermal rectifier is a device that causes heat to flow preferentially in one direction.

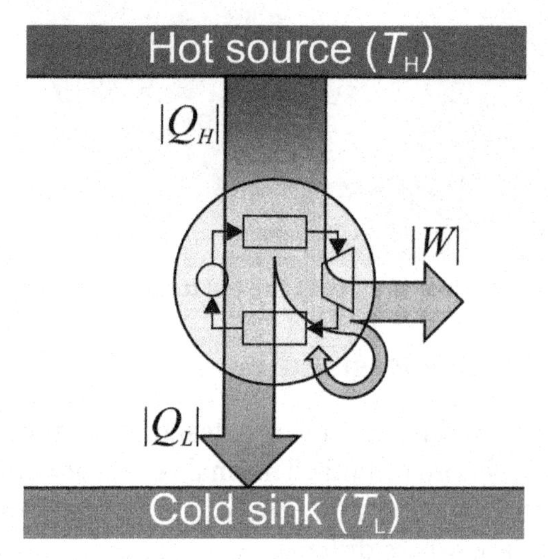

Heat engine diagram

Heat Exchangers

A heat exchanger is used for more efficient heat transfer or to dissipate heat. Heat exchangers are widely used in refrigeration, air conditioning, space heating, power generation, and chemical processing. One common example of a heat exchanger is a car's radiator, in which the hot coolant fluid is cooled by the flow of air over the radiator's surface.

Common types of heat exchanger flows include parallel flow, counter flow, and cross flow. In parallel flow, both fluids move in the same direction while transferring heat; in counter flow, the fluids move in opposite directions; and in cross flow, the fluids move at right angles to each other. Common constructions for heat exchanger include shell and tube, double pipe, extruded finned pipe, spiral fin pipe, u-tube, and stacked plate.

A heat sink is a component that transfers heat generated within a solid material to a fluid medium, such as air or a liquid. Examples of heat sinks are the heat exchangers used in refrigeration and air conditioning systems or the radiator in a car. A heat pipe is another heat-transfer device that combines thermal conductivity and phase transition to efficiently transfer heat between two solid interfaces.

Examples

Architecture

Efficient energy use is the goal to reduce the amount of energy required in heating or cooling. In architecture, condensation and air currents can cause cosmetic or structural damage. An energy audit can help to assess the implementation of recommended corrective procedures. For instance, insulation improvements, air sealing of structural leaks or the addition of energy-efficient windows and doors.

- Smart meter is a device that records electric energy consumption in intervals.
- Thermal transmittance is the rate of transfer of heat through a structure divided by the difference in temperature across the structure. It is expressed in watts per square meter per kelvin, or W/m^2K. Well-insulated parts of a building have a low thermal transmittance, whereas poorly-insulated parts of a building have a high thermal transmittance.
- Thermostat is a device to monitor and control temperature.

Climate Engineering

An example application in climate engineering includes the creation of Biochar through the pyrolysis process. Thus, storing greenhouse gases in carbon reduces the radiative forcing capacity in the atmosphere, causing more long-wave (infrared) radiation out to Space.

Climate engineering consists of carbon dioxide removal and solar radiation management. Since the amount of carbon dioxide determines the radiative balance of Earth atmosphere, carbon dioxide removal techniques can be applied to reduce the radiative forcing. Solar radiation management is the attempt to absorb less solar radiation to offset the effects of greenhouse gases.

Greenhouse Effect

The greenhouse effect is a process by which thermal radiation from a planetary surface is absorbed by atmospheric greenhouse gases, and is re-radiated in all directions. Since part of this re-radiation is back towards the surface and the lower atmosphere, it results in an elevation of the average surface temperature above what it would be in the absence of the gases.

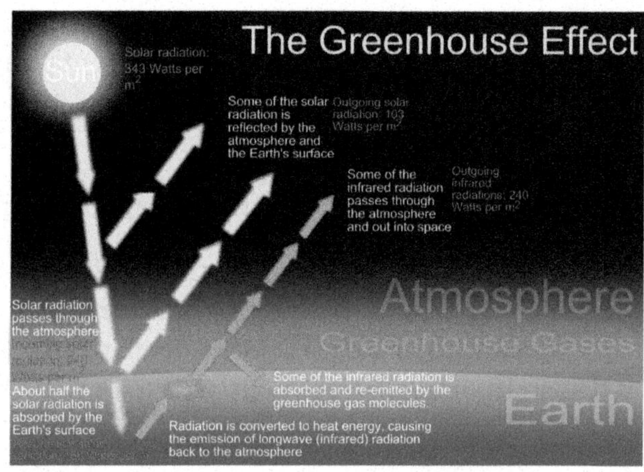

A representation of the exchanges of energy between the source (the Sun), the Earth's surface, the Earth's atmosphere, and the ultimate sink outer space. The ability of the atmosphere to capture and recycle energy emitted by the Earth surface is the defining characteristic of the greenhouse effect.

Heat Transfer in the Human Body

The principles of heat transfer in engineering systems can be applied to the human body in order to determine how the body transfers heat. Heat is produced in the body by the continuous metabolism of nutrients which provides energy for the systems of the body. The human body must maintain a consistent internal temperature in order to maintain healthy bodily functions. Therefore, excess heat must be dissipated from the body to keep it from overheating. When a person engages in elevated levels of physical activity, the body requires additional fuel which increases the metabolic rate and the rate of heat production. The body must then use additional methods to remove the additional heat produced in order to keep the internal temperature at a healthy level.

Heat transfer by convection is driven by the movement of fluids over the surface of the body. This convective fluid can be either a liquid or a gas. For heat transfer from the outer surface of the body, the convection mechanism is dependent on the surface area of the body, the velocity of the air, and the temperature gradient between the surface of the skin and the ambient air. The normal temperature of the body is approximately 37 °C. Heat transfer occurs more readily when the temperature of the surroundings is significantly less than the normal body temperature. This concept explains why a person feels "cold" when not enough covering is worn when exposed to a cold environment.

Clothing can be considered an insulator which provides thermal resistance to heat flow over the covered portion of the body. This thermal resistance causes the temperature on the surface of the clothing to be less than the temperature on the surface of the skin. This smaller temperature gradient between the surface temperature and the ambient temperature will cause a lower rate of heat transfer than if the skin were not covered.

In order to ensure that one portion of the body is not significantly hotter than another portion, heat must be distributed evenly through the bodily tissues. Blood flowing through blood vessels acts as a convective fluid and helps to prevent any buildup of excess heat inside the tissues of the body. This flow of blood through the vessels can be modeled as pipe flow in an engineering system. The heat carried by the blood is determined by the temperature of the surrounding tissue, the diameter of the blood vessel, the thickness of the fluid, velocity of the flow, and the heat transfer coefficient of the blood. The velocity, blood vessel diameter, and the fluid thickness can all be related with the Reynolds Number, a dimensionless number used in fluid mechanics to characterize the flow of fluids.

Latent heat loss, also known as evaporative heat loss, accounts for a large fraction of heat loss from the body. When the core temperature of the body increases, the body triggers sweat glands in the skin to bring additional moisture to the surface of the skin. The liquid is then transformed into vapor which removes heat from the surface of the body. The rate of evaporation heat loss is directly related to the vapor pressure at the skin surface and the amount of moisture present on the skin. Therefore, the maximum of heat transfer will occur when the skin is completely wet. The body continuously loses water by evaporation but the most significant amount of heat loss occurs during periods of increased physical activity.

Cooling Techniques

Evaporative Cooling

A traditional air cooler in Mirzapur, Uttar Pradesh, India

Evaporative cooling happens when water vapor is added to the surrounding air. The energy needed to evaporate the water is taken from the air in the form of sensible heat and converted into latent heat, while the air remains at a constant enthalpy. Latent heat describes the amount of heat that is needed to evaporate the liquid; this heat comes from the liquid itself and the surrounding gas and surfaces. The greater the difference between the two temperatures, the greater the evaporative cooling effect. When the temperatures are the same, no net evaporation of water in air occurs; thus, there is no cooling effect.

Laser Cooling

In Quantum Physics laser cooling is used to achieve temperatures of near absolute zero (−273.15 °C, −459.67 °F) of atomic and molecular samples, to observe unique quantum effects that can only occur at this heat level.

- Doppler cooling is the most common method of laser cooling.
- Sympathetic cooling is a process in which particles of one type cool particles of another type. Typically, atomic ions that can be directly laser-cooled are used to cool nearby ions or atoms. This technique allows cooling of ions and atoms that cannot be laser cooled directly.

Magnetic Cooling

Magnetic evaporative cooling is a process for lowering the temperature of a group of atoms, after pre-cooled by methods such as laser cooling. Magnetic refrigeration cools below 0.3K, by making use of the magnetocaloric effect.

Radiative Cooling

Radiative cooling is the process by which a body loses heat by radiation. Outgoing energy is an important effect in the Earth's energy budget. In the case of the Earth-atmosphere system, it refers to the process by which long-wave (infrared) radiation is emitted to balance the absorption of short-wave (visible) energy from the Sun. Convective transport of heat and evaporative transport of latent heat both remove heat from the surface and redistribute it in the atmosphere.

Thermal Energy Storage

Thermal energy storage refers to technologies used to collect and store energy for later use. They can be employed to balance energy demand between day and nighttime. The thermal reservoir may be maintained at a temperature above (hotter) or below (colder) than that of the ambient environment. Applications include later use in space heating, domestic or process hot water, or to generate electricity.

References

- Lienhard, John H.,V; Lienhard, John H., V (2008). A Heat Transfer Textbook (3rd ed.). Cambridge, Massachusetts: Phlogiston Press. ISBN 978-0-9713835-3-1. OCLC 230956959.
- Welty, James R.; Wicks, Charles E.; Wilson, Robert Elliott (1976). Fundamentals of momentum, heat, and mass transfer (2 ed.). New York: Wiley. ISBN 978-0-471-93354-0. OCLC 2213384.
- Faghri, Amir; Zhang, Yuwen; Howell, John (2010). Advanced Heat and Mass Transfer. Columbia, MO: Global Digital Press. ISBN 978-0-9842760-0-4.
- Paul A., Tipler; Gene Mosca (2008). Physics for Scientists and Engineers, Volume 1 (6th ed.). New York, NY: Worth Publishers. pp. 666–670. ISBN 1-4292-0132-0.
- Abbott, J.M. Smith, H.C. Van Ness, M.M. (2005). Introduction to chemical engineering thermodynamics (7th ed.). Boston ; Montreal: McGraw-Hill. ISBN 0-07-310445-0.
- Çengel, Yunus (2003). Heat Transfer: a practical approach. McGraw-Hill series in mechanical engineering. (2nd ed.). Boston: McGraw-Hill. ISBN 978-0-07-245893-0. OCLC 300472921. Retrieved 2009-04-20.
- Geankoplis, Christie John (2003). Transport processes and separation process principles : (includes unit operations) (4th ed.). Upper Saddle River, NJ: Prentice Hall Professional Technical Reference. ISBN 0-13-101367-X.
- "EnergySavers: Tips on Saving Money & Energy at Home" (PDF). U.S. Department of Energy. Retrieved March 2, 2012.
- New Jersey Institute of Technology, Chemical Engineering Dept. "B.S. Chemical Engineering". NJIT. Retrieved 9 April 2011.

2

Heat: An Overview

The energy that is passed between a system and its surroundings is termed as heat; heat usually transfers from a hotter to a colder body. The following chapter will not only provide an overview, it will also examine into the variegated topics related to it.

In physics, heat is energy that spontaneously passes between a system and its surroundings in some way other than through work or the transfer of matter. When a suitable physical pathway exists, heat flows spontaneously from a hotter to a colder body. The transfer can be by contact between the source and the destination body, as in conduction; or by radiation between remote bodies; or by conduction and radiation through a thick solid wall; or by way of an intermediate fluid body, as in convective circulation; or by a combination of these.

Because heat refers to a quantity of energy transferred between two bodies, it is not a state function of either of the bodies, in contrast to temperature and internal energy. Instead, according to the first law of thermodynamics heat exchanged during some process contributes to the change in the internal energy, and the amount of heat can be quantified by the equivalent amount of work that would bring about the same change.

While heat flows spontaneously from hot to cold, it is possible to construct a heat pump or refrigeration system that does work to increase the difference in temperature between two systems. Conversely, a heat engine reduces an existing temperature difference to do work on another system.

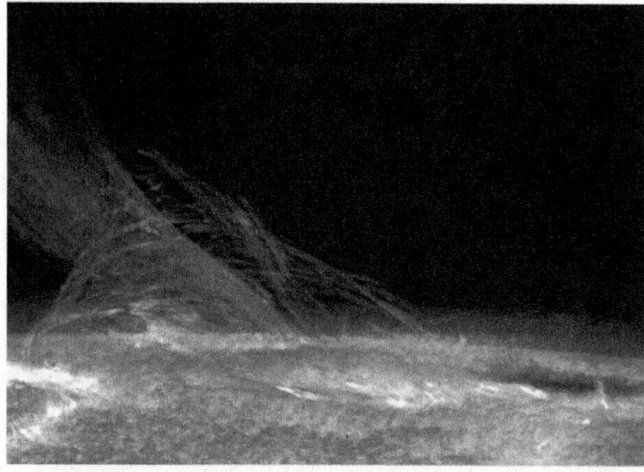

The Sun and Earth form an ongoing example of a heating process. Some of the Sun's thermal radiation strikes and heats the Earth. Compared to the Sun, Earth has a much lower temperature and so sends far less thermal radiation back to the Sun. The heat of this process can be quantified by the net amount, and direction (Sun to Earth), of energy it transferred in a given period of time.

Historically, many energy units for measurement of heat have been used. The standards-based unit in the International System of Units (SI) is the joule (J). Heat is measured by its effect on the states of interacting bodies, for example, by the amount of ice melted or a change in temperature. The quantification of heat via the temperature change of a body is called calorimetry, and is widely used in practice. In calorimetry, sensible heat is defined with respect to a specific chosen state variable of the system, such as pressure or volume. Sensible heat causes a change of the temperature of the system while leaving the chosen state variable unchanged. Heat transfer that occurs at a constant system temperature but changes the state variable is called latent heat with respect to the variable. For infinitesimal changes, the total incremental heat transfer is then the sum of the latent and sensible heat.

History

Physicist James Clerk Maxwell, in his 1871 classic *Theory of Heat,* was one of many who began to build on the already established idea that heat has something to do with matter in motion. This was the same idea put forth by Benjamin Thompson in 1798, who said he was only following up on the work of many others. One of Maxwell's recommended books was *Heat as a Mode of Motion*, by John Tyndall. Maxwell outlined four stipulations for the definition of heat:

- It is *something which may be transferred from one body to another*, according to the second law of thermodynamics.
- It is a *measurable quantity*, and so can be treated mathematically.
- It *cannot be treated as a material substance*, because it may be transformed into something that is not a material substance, e.g., mechanical work.
- Heat is *one of the forms of energy.*

From empirically based ideas of heat, and from other empirical observations, the notions of internal energy and of entropy can be derived, so as to lead to the recognition of the first and second laws of thermodynamics. This was the way of the historical pioneers of thermodynamics.

Transfers of Energy as Heat Between Two Bodies

Referring to conduction, Partington writes: "If a hot body is brought in conducting contact with a cold body, the temperature of the hot body falls and that of the cold body rises, and it is said that a *quantity of heat* has passed from the hot body to the cold body."

Referring to radiation, Maxwell writes: "In Radiation, the hotter body loses heat, and the colder body receives heat by means of a process occurring in some intervening medium which does not itself thereby become hot."

Maxwell writes that convection as such "is not a purely thermal phenomenon". In thermodynamics, convection in general is regarded as transport of internal energy. If, however, the convection is enclosed and circulatory, then it may be regarded as an intermediary that transfers energy as heat between source and destination bodies, because it transfers only energy and not matter from the source to the destination body.

Practical Operating Devices that Harness Transfers of Energy as Heat

In accordance with the first law for closed systems, energy transferred solely as heat leaves one body and enters another, changing the internal energies of each. Transfer, between bodies, of energy as work is a complementary way of changing internal energies. Though it is not logically rigorous from the viewpoint of strict physical concepts, a common form of words that expresses this is to say that heat and work are interconvertible.

Cyclically operating engines, that use only heat and work transfers, have two thermal reservoirs, a hot and a cold one. They may be classified by the range of operating temperatures of the working body, relative to those reservoirs. In a heat engine, the working body is at all times colder than the hot reservoir and hotter than the cold reservoir. In a sense, it uses heat transfer to produce work. In a heat pump, the working body, at stages of the cycle, goes both hotter than the hot reservoir, and colder than the cold reservoir. In a sense, it uses work to produce heat transfer.

Heat Engine

In classical thermodynamics, a commonly considered model is the heat engine. It consists of four bodies: the working body, the hot reservoir, the cold reservoir, and the work reservoir. A cyclic process leaves the working body in an unchanged state, and is envisaged as being repeated indefinitely often. Work transfers between the working body and the work reservoir are envisaged as reversible, and thus only one work reservoir is needed. But two thermal reservoirs are needed, because transfer of energy as heat is irreversible. A single cycle sees energy taken by the working body from the hot reservoir and sent to the two other reservoirs, the work reservoir and the cold reservoir. The hot reservoir always and only supplies energy and the cold reservoir always and only receives energy. The second law of thermodynamics requires that no cycle can occur in which no energy is received by the cold reservoir. Heat engines achieve higher efficiency when the difference between initial and final temperature is greater.

Heat Pump or Refrigerator

Another commonly considered model is the heat pump or refrigerator. Again there are four bodies: the working body, the hot reservoir, the cold reservoir, and the work reservoir. A single cycle starts with the working body colder than the cold reservoir, and then energy is taken in as heat by the working body from the cold reservoir. Then the work reservoir does work on the working body, adding more to its internal energy, making it hotter than the hot reservoir. The hot working body passes heat to the hot reservoir, but still remains hotter than the cold reservoir. Then, by allowing it to expand without doing work on another body and without passing heat to another body, the working body is made colder than the cold reservoir. It can now accept heat transfer from the cold reservoir to start another cycle.

The device has transported energy from a colder to a hotter reservoir, but this is not regarded as being by an inanimate agency; rather, it is regarded as by the harnessing of work . This is because work is supplied from the work reservoir, not just by a simple thermodynamic process, but by a cycle of thermodynamic operations and processes, which may be regarded as directed by an animate or harnessing agency. Accordingly, the cycle is still in accord with the second law of thermodynamics. The efficiency of a heat pump is best when the temperature difference between the hot and cold reservoirs is least.

Functionally, such engines are used in two ways, distinguishing a target reservoir and a resource or surrounding reservoir. A heat pump transfers heat, to the hot reservoir as the target, from the resource or surrounding reservoir. A refrigerator transfers heat, from the cold reservoir as the target, to the resource or surrounding reservoir. The target reservoir may be regarded as leaking: when the target leaks hotness to the surroundings, heat pumping is used; when the target leaks coldness to the surroundings, refrigeration is used. The engines harness work to overcome the leaks.

Macroscopic View of Quantity of Energy Transferred as Heat

According to Planck, there are three main conceptual approaches to heat. One is the microscopic or kinetic theory approach. The other two are macroscopic approaches. One is the approach through the law of conservation of energy taken as prior to thermodynamics, with a mechanical analysis of processes, for example in the work of Helmholtz. This mechanical view is taken in this article as currently customary for thermodynamic theory. The other macroscopic approach is the thermodynamic one, which admits heat as a primitive concept, which contributes, by scientific induction to knowledge of the law of conservation of energy. This view is widely taken as the practical one, quantity of heat being measured by calorimetry.

Bailyn also distinguishes the two macroscopic approaches as the mechanical and the thermodynamic. The thermodynamic view was taken by the founders of thermodynamics in the nineteenth century. It regards quantity of energy transferred as heat as a primitive concept coherent with a primitive concept of temperature, measured primarily by calorimetry. A calorimeter is a body in the surroundings of the system, with its own temperature and internal energy; when it is connected to the system by a path for heat transfer, changes in it measure heat transfer. The mechanical view was pioneered by Helmholtz and developed and used in the twentieth century, largely through the influence of Max Born. It regards quantity of heat transferred as heat as a derived concept, defined for closed systems as quantity of heat transferred by mechanisms other than work transfer, the latter being regarded as primitive for thermodynamics, defined by macroscopic mechanics. According to Born, the transfer of internal energy between open systems that accompanies transfer of matter "cannot be reduced to mechanics". It follows that there is no well-founded definition of quantities of energy transferred as heat or as work associated with transfer of matter.

Nevertheless, for the thermodynamical description of non-equilibrium processes, it is desired to consider the effect of a temperature gradient established by the surroundings across the system of interest when there is no physical barrier or wall between system and surroundings, that is to say, when they are open with respect to one another. The impossibility of a mechanical definition in terms of work for this circumstance does not alter the physical fact that a temperature gradient causes a diffusive flux of internal energy, a process that, in the thermodynamic view, might be proposed as a candidate concept for transfer of energy as heat.

In this circumstance, it may be expected that there may also be active other drivers of diffusive flux of internal energy, such as gradient of chemical potential which drives transfer of matter, and gradient of electric potential which drives electric current and iontophoresis; such effects usually interact with diffusive flux of internal energy driven by temperature gradient, and such interactions are known as cross-effects.

If cross-effects that result in diffusive transfer of internal energy were also labeled as heat transfers, they would sometimes violate the rule that pure heat transfer occurs only down a temperature gradient, never up one. They would also contradict the principle that all heat transfer is of one and the same kind, a principle founded on the idea of heat conduction between closed systems. One might to try to think narrowly of heat flux driven purely by temperature gradient as a conceptual component of diffusive internal energy flux, in the thermodynamic view, the concept resting specifically on careful calculations based on detailed knowledge of the processes and being indirectly assessed. In these circumstances, if perchance it happens that no transfer of matter is actualized, and there are no cross-effects, then the thermodynamic concept and the mechanical concept coincide, as if one were dealing with closed systems. But when there is transfer of matter, the exact laws by which temperature gradient drives diffusive flux of internal energy, rather than being exactly knowable, mostly need to be assumed, and in many cases are practically unverifiable. Consequently, when there is transfer of matter, the calculation of the pure 'heat flux' component of the diffusive flux of internal energy rests on practically unverifiable assumptions. This is a reason to think of heat as a specialized concept that relates primarily and precisely to closed systems, and applicable only in a very restricted way to open systems.

In many writings in this context, the term "heat flux" is used when what is meant is therefore more accurately called diffusive flux of internal energy; such usage of the term "heat flux" is a residue of older and now obsolete language usage that allowed that a body may have a "heat content".

Microscopic View of Heat

In the kinetic theory, heat is explained in terms of the microscopic motions and interactions of constituent particles, such as electrons, atoms, and molecules. The immediate meaning of the kinetic energy of the constituent particles is not as heat. It is as a component of internal energy. In microscopic terms, heat is a transfer quantity, and is described by a transport theory, not as steadily localized kinetic energy of particles. Heat transfer arises from temperature gradients or differences, through the diffuse exchange of microscopic kinetic and potential particle energy, by particle collisions and other interactions. An early and vague expression of this was made by Francis Bacon. Precise and detailed versions of it were developed in the nineteenth century.

In statistical mechanics, for a closed system (no transfer of matter), heat is the energy transfer associated with a disordered, microscopic action on the system, associated with jumps in occupation numbers of the energy levels of the system, without change in the values of the energy levels themselves. It is possible for macroscopic thermodynamic work to alter the occupation numbers without change in the values of the system energy levels themselves, but what distinguishes transfer as heat is that the transfer is entirely due to disordered, microscopic action, including radiative transfer. A mathematical definition can be formulated for small increments of quasi-static adiabatic work in terms of the statistical distribution of an ensemble of microstates.

Notation and Units

As a form of energy heat has the unit joule (J) in the International System of Units (SI). However, in many applied fields in engineering the British thermal unit (BTU) and the calorie are often used. The standard unit for the rate of heat transferred is the watt (W), defined as joules per second.

The total amount of energy transferred as heat is conventionally written as Q (from Quantity) for algebraic purposes. Heat released by a system into its surroundings is by convention a negative quantity ($Q < 0$); when a system absorbs heat from its surroundings, it is positive ($Q > 0$). Heat transfer rate, or heat flow per unit time, is denoted by . This should not be confused with a time derivative of a function of state (which can also be written with the dot notation) since heat is not a function of state. Heat flux is defined as rate of heat transfer per unit cross-sectional area, resulting in the unit *watts per square metre*.

Estimation of Quantity of Heat

Quantity of heat transferred can measured by calorimetry, or determined through calculations based on other quantities.

Calorimetry is the empirical basis of the idea of quantity of heat transferred in a process. The transferred heat is measured by changes in a body of known properties, for example, temperature rise, change in volume or length, or phase change, such as melting of ice.

A calculation of quantity of heat transferred can rely on a hypothetical quantity of energy transferred as adiabatic work and on the first law of thermodynamics. Such calculation is the primary approach of many theoretical studies of quantity of heat transferred.

Internal Energy and Enthalpy

For a closed system (a system from which no matter can enter or exit), one version of the first law of thermodynamics states that the change in internal energy ΔU of the system is equal to the amount of heat Q supplied to the system minus the amount of work W done by system on its surroundings. The foregoing sign convention for work is used in the present article, but an alternate sign convention, followed by IUPAC, for work, is to consider the work performed on the system by its surroundings as positive. This is the convention adopted by many modern textbooks of physical chemistry, such as those by Peter Atkins and Ira Levine, but many textbooks on physics define work as work done by the system.

$$\Delta U = Q - W.$$

This formula can be re-written so as to express a definition of quantity of energy transferred as heat, based purely on the concept of adiabatic work, if it is supposed that ΔU is defined and measured solely by processes of adiabatic work:

$$Q = \Delta U + W.$$

The work done by the system includes boundary work (when the system increases its volume against an external force, such as that exerted by a piston) and other work (e.g. shaft work performed by a compressor fan), which is called isochoric work:

$$Q = \Delta U + W_{\text{boundary}} + W_{\text{isochoric}}.$$

In this Section we will neglect the "other-" or isochoric work contribution.

The internal energy, U, is a state function. In cyclical processes, such as the operation of a heat engine, state functions of the working substance return to their initial values upon completion of a cycle.

The differential, or infinitesimal increment, for the internal energy in an infinitesimal process is an exact differential dU. The symbol for exact differentials is the lowercase letter d.

In contrast, neither of the infinitesimal increments δQ nor δW in an infinitesimal process represents the state of the system. Thus, infinitesimal increments of heat and work are inexact differentials. The lowercase Greek letter delta, δ, is the symbol for inexact differentials. The integral of any inexact differential over the time it takes for a system to leave and return to the same thermodynamic state does not necessarily equal zero.

As recounted below, in the section headed Entropy, the second law of thermodynamics observes that if heat is supplied to a system in which no irreversible processes take place and which has a well-defined temperature T, the increment of heat δQ and the temperature T form the exact differential

$$\mathrm{d}S = \frac{\delta Q}{T},$$

and that S, the entropy of the working body, is a function of state. Likewise, with a well-defined pressure, P, behind the moving boundary, the work differential, δW, and the pressure, P, combine to form the exact differential

$$\mathrm{d}V = \frac{\delta W}{P},$$

with V the volume of the system, which is a state variable. In general, for homogeneous systems

$$\mathrm{d}U = T\mathrm{d}S - P\mathrm{d}V.$$

Associated with this differential equation is that the internal energy may be considered to be a function $U(S,V)$ of its natural variables S and V. The internal energy representation of the fundamental thermodynamic relation is written

$$U = U(S,V).$$

If V is constant

$$T\mathrm{d}S = \mathrm{d}U \quad (V \text{ constant})$$

and if P is constant

$$T\mathrm{d}S = \mathrm{d}H \quad (P \text{ constant})$$

with H the enthalpy defined by

$$H = U + PV.$$

The enthalpy may be considered to be a function $H(S,P)$ of its natural variables S and P. The enthalpy representation of the fundamental thermodynamic relation is written

$$H = H(S,P).$$

The internal energy representation and the enthalpy representation are partial Legendre transforms of one another. They contain the same physical information, written in different ways. Like the internal energy, the enthalpy stated as a function of its natural variables is a thermodynamic potential and contains all thermodynamic information about a body.

Heat Added to a Body at Constant Pressure

If a quantity Q of heat is added to a body while it does expansion work W on its surroundings, one has

$$\Delta H = \Delta U + \Delta(PV).$$

If this is constrained to happen at constant pressure with $\Delta P = 0$, the expansion work W done by the body is given by $W = P\,\Delta V$; recalling the first law of thermodynamics, one has

$$\Delta U = Q - W = Q - P\Delta V \text{ and } \Delta(PV) = P\Delta V.$$

Consequently, by substitution one has

$$\Delta H = Q - P\Delta V + P\Delta V$$

$$= Q \qquad \text{at constant pressure.}$$

In this scenario, the increase in enthalpy is equal to the quantity of heat added to the system. Since many processes do take place at constant pressure, or approximately at atmospheric pressure, the enthalpy is therefore sometimes given the misleading name of 'heat content'. It is sometimes also called the heat function.

In terms of the natural variables S and P of the state function H, this process of change of state from state 1 to state 2 can be expressed as

$$\Delta H = \int_{S_1}^{S_2} \left(\frac{\partial H}{\partial S}\right)_P \mathrm{d}S + \int_{P_1}^{P_2} \left(\frac{\partial H}{\partial P}\right)_S \mathrm{d}P$$

$$= \int_{S_1}^{S_2} \left(\frac{\partial H}{\partial S}\right)_P \mathrm{d}S \qquad \text{at constant pressure.}$$

It is known that the temperature $T(S, P)$ is identically stated by

$$\left(\frac{\partial H}{\partial S}\right)_P \equiv T(S,P).$$

Consequently,

$$\Delta H = \int_{S_1}^{S_2} T(S,P)\mathrm{d}S \qquad \text{at constant pressure.}$$

In this case, the integral specifies a quantity of heat transferred at constant pressure.

Entropy

Rudolf Clausius

In 1856, German physicist Rudolf Clausius, referring to closed systems, in which transfers of matter do not occur, defined the *second fundamental theorem* (the second law of thermodynamics) in the mechanical theory of heat (thermodynamics): "if two transformations which, without necessitating any other permanent change, can mutually replace one another, be called equivalent, then the generations of the quantity of heat Q from work at the temperature T, has the *equivalence-value*:"

$$\frac{Q}{T}.$$

In 1865, he came to define the entropy symbolized by S, such that, due to the supply of the amount of heat Q at temperature T the entropy of the system is increased by

$$\Delta S = \frac{Q}{T} \quad (1)$$

In a transfer of energy as heat without work being done, there are changes of entropy in both the surroundings which lose heat and the system which gains it. The increase, ΔS, of entropy in the system may be considered to consist of two parts, an increment, $\Delta S'$ that matches, or 'compensates', the change, $-\Delta S'$, of entropy in the surroundings, and a further increment, $\Delta S''$ that may be considered to be 'generated' or 'produced' in the system, and is said therefore to be 'uncompensated'. Thus

$$\Delta S = \Delta S' + \Delta S''.$$

This may also be written

$$\Delta S_{\text{system}} = \Delta S_{\text{compensated}} + \Delta S_{\text{uncompensated}} \text{ with } \Delta S_{\text{compensated}} = -\Delta S_{\text{surroundings}}.$$

The total change of entropy in the system and surroundings is thus

$$\Delta S_{\text{overall}} = \Delta S' + \Delta S'' - \Delta S' = \Delta S''.$$

This may also be written

$$\Delta S_{\text{overall}} = \Delta S_{\text{compensated}} + \Delta S_{\text{uncompensated}} + \Delta S_{\text{surroundings}} = \Delta S_{\text{uncompensated}}.$$

It is then said that an amount of entropy $\Delta S'$ has been transferred from the surroundings to the system. Because entropy is not a conserved quantity, this is an exception to the general way of speaking, in which an amount transferred is of a conserved quantity.

The second law of thermodynamics observes that in a natural transfer of energy as heat, in which the temperature of the system is different from that of the surroundings, it is always so that

$$\Delta S_{\text{overall}} > 0.$$

For purposes of mathematical analysis of transfers, one thinks of fictive processes that are called 'reversible', with the temperature T of the system being hardly less than that of the surroundings, and the transfer taking place at an imperceptibly slow speed.

Following the definition above in formula (1), for such a fictive 'reversible' process, a quantity of transferred heat δQ (an inexact differential) is analyzed as a quantity $T\,\mathrm{d}S$, with $\mathrm{d}S$ (an exact differential):

$$T\,\mathrm{d}S = \delta Q.$$

This equality is only valid for a fictive transfer in which there is no production of entropy, that is to say, in which there is no uncompensated entropy.

If, in contrast, the process is natural, and can really occur, with irreversibility, then there is entropy production, with $\mathrm{d}S_{\text{uncompensated}} > 0$. The quantity $T\,\mathrm{d}S_{\text{uncompensated}}$ was termed by Clausius the "uncompensated heat", though that does not accord with present-day terminology. Then one has

$$T\,\mathrm{d}S = \delta Q + T\,\mathrm{d}S_{\text{uncompensated}} > \delta Q.$$

This leads to the statement

$$T\,\mathrm{d}S \geq \delta Q \quad \text{(second law)}.$$

which is the second law of thermodynamics for closed systems.

In non-equilibrium thermodynamics that approximates by assuming the hypothesis of local thermodynamic equilibrium, there is a special notation for this. The transfer of energy as heat is assumed to take place across an infinitesimal temperature difference, so that the system element and its surroundings have near enough the same temperature T. Then one writes

$$\mathrm{d}S = \mathrm{d}S_{\mathrm{e}} + \mathrm{d}S_{\mathrm{i}},$$

where by definition

$$\delta Q = T\,\mathrm{d}S_{\mathrm{e}} \text{ and } \mathrm{d}S_{\mathrm{i}} \equiv \mathrm{d}S_{\mathrm{uncompensated}}.$$

The second law for a natural process asserts that

$$\mathrm{d}S_{\mathrm{i}} > 0.$$

Latent and Sensible Heat

In an 1847 lecture entitled *On Matter, Living Force, and Heat*, James Prescott Joule characterized the terms latent heat and sensible heat as components of heat each affecting distinct physical phenomena, namely the potential and kinetic energy of particles, respectively.He described latent energy as the energy possessed via a distancing of particles where attraction was over a greater distance, i.e. a form of potential energy, and the sensible heat as an energy involving the motion of particles or what was known as a *living force*. At the time of Joule kinetic energy either held 'invisibly' internally or held 'visibly' externally was known as a *living force*.

Joseph Black

Latent heat is the heat released or absorbed by a chemical substance or a thermodynamic system during a change of state that occurs without a change in temperature. Such a process may be a phase transition, such as the melting of ice or the boiling of water. The term was introduced around 1750 by Joseph Black as derived from the Latin *latere* (*to lie hidden*), characterizing its effect as not being directly measurable with a thermometer.

Sensible heat, in contrast to latent heat, is the heat transferred to a thermodynamic system that has as its sole effect a change of temperature.

Both latent heat and sensible heat transfers increase the internal energy of the system to which they are transferred.

Consequences of Black's distinction between sensible and latent heat are examined in the article on calorimetry.

Specific Heat

Specific heat, also called specific heat capacity, is defined as the amount of energy that has to be transferred to or from one unit of mass (kilogram) or amount of substance (mole) to change the system temperature by one degree. Specific heat is a physical property, which means that it depends on the substance under consideration and its state as specified by its properties.

The specific heats of monatomic gases (e.g., helium) are nearly constant with temperature. Diatomic gases such as hydrogen display some temperature dependence, and triatomic gases (e.g., carbon dioxide) still more.

Relation Between Heat and Temperature

Before the discovery of the laws of thermodynamics, quantity of energy transferred as heat was measured by changes in the states of the participating bodies.

Some general rules, with important exceptions, that will be indicated noted in following paragraphs of this section, can be stated as follows.

Most bodies, over most temperature ranges, expand on being heated. Mostly, heating a body at a constant volume increases the pressure it exerts on its constraining walls, and increases its temperature. Also mostly, heating a body at a constant pressure increases its volume, and increases its temperature.

Beyond this, most substances have three ordinarily recognized states of matter, solid, liquid, and gas, and a fourth less obviously recognized one, plasma. Many have further, more finely differentiated, states of matter, such as for example, glass, and liquid crystal. In many cases, at fixed temperature and pressure, a substance can exist in several distinct states of matter in what might be viewed as the same 'body'. For example, ice may float in a glass of water. Then the ice and the water are said to constitute two phases within the 'body'. Definite rules are known, telling how distinct phases may coexist in a 'body'. Mostly, at a fixed pressure, there is a definite temperature at which heating causes a solid to melt or evaporate, and a definite temperature at which heating causes a liquid to evaporate. In such cases, cooling has the reverse effects.

All of these, the commonest cases, fit with a rule that heating can be measured by changes of state of a body. Such cases supply what are called *thermometric bodies*, that allow the definition of empirical temperatures. Before 1848, all temperatures were defined in this way. There was thus a tight link, apparently logically determined, between heat and temperature, though they were recognized as conceptually thoroughly distinct, especially by Joseph Black in the later eighteenth century.

There are important exceptions. They break the obviously apparent link between heat and temperature. They make it clear that empirical definitions of temperature are contingent on the peculiar properties of particular thermometric substances, and are thus precluded from the title 'absolute'. For example, water contracts on being heated near 277 K. It cannot be used as a thermometric substance near that temperature. Also, over a certain temperature range, ice contracts on heating. Moreover, many substances can exist in metastable states, such as with negative pressure, that survive only transiently and in very special conditions. Such facts, sometimes called 'anomalous', are some of the reasons for the thermodynamic definition of absolute temperature.

In the early days of measurement of high temperatures, another factor was important, and used by Josiah Wedgwood in his pyrometer. The temperature reached in a process was estimated by the shrinkage of a sample of clay. The higher the temperature, the more the shrinkage. This was the only available more or less reliable method of measurement of temperatures above 1000 °C. But such shrinkage is irreversible. The clay does not expand again on cooling. That is why it could be used for the measurement. But only once. It is not a thermometric material in the usual sense of the word.

Nevertheless, the thermodynamic definition of absolute temperature does make essential use of the concept of heat, with proper circumspection.

Relation Between Hotness and Temperature

According to Denbigh, the property of hotness is a concern of thermodynamics that should be defined without reference to the concept of heat. Consideration of hotness leads to the concept of empirical temperature. All physical systems are capable of heating or cooling others. This does not require that they have thermodynamic temperatures. With reference to hotness, the comparative terms hotter and colder are defined by the rule that heat flows from the hotter body to the colder.

If a physical system is inhomogeneous or very rapidly or irregularly changing, for example by turbulence, it may be impossible to characterize it by a temperature, but still there can be transfer of energy as heat between it and another system. If a system has a physical state that is regular enough, and persists long enough to allow it to reach thermal equilibrium with a specified thermometer, then it has a temperature according to that thermometer. An empirical thermometer registers degree of hotness for such a system. Such a temperature is called empirical. For example, Truesdell writes about classical thermodynamics: "At each time, the body is assigned a real number called the *temperature*. This number is a measure of how hot the body is."

Physical systems that are too turbulent to have temperatures may still differ in hotness. A physical system that passes heat to another physical system is said to be the hotter of the two. More is required for the system to have a thermodynamic temperature. Its behavior must be so regular that its empirical temperature is the same for all suitably calibrated and scaled thermometers, and then its hotness is said to lie on the one-dimensional hotness manifold. This is part of the reason why heat is defined following Carathéodory and Born, solely as occurring other than by work or transfer of matter; temperature is advisedly and deliberately not mentioned in this now widely accepted definition.

This is also the reason why the zeroth law of thermodynamics is stated explicitly. If three physical systems, *A*, *B*, and *C* are each not in their own states of internal thermodynamic equilibrium, it is possible that, with suitable physical connections being made between them, *A* can heat *B* and *B* can heat *C* and *C* can heat *A*. In non-equilibrium situations, cycles of flow are possible. It is the special and uniquely distinguishing characteristic of internal thermodynamic equilibrium that this possibility is not open to thermodynamic systems (as distinguished amongst physical systems) which are in their own states of internal thermodynamic equilibrium; this is the reason why the zeroth law of thermodynamics needs explicit statement. That is to say, the relation 'is not colder than' between general non-equilibrium physical systems is not transitive, whereas, in contrast, the relation 'has no lower a temperature than' between thermodynamic systems in their own states of internal thermodynamic equilibrium is transitive. It follows from this that the relation 'is in thermal equilibrium with' is transitive, which is one way of stating the zeroth law.

Just as temperature may undefined for a sufficiently inhomogeneous system, so also may entropy be undefined for a system not in its own state of internal thermodynamic equilibrium. For example, 'the temperature of the solar system' is not a defined quantity. Likewise, 'the entropy of the solar system' is not defined in classical thermodynamics. It has not been possible to define non-equilibrium entropy, as a simple number for a whole system, in a clearly satisfactory way.

Rigorous Definition of Quantity of Energy Transferred as Heat

It is sometimes convenient to have a rigorous definition of quantity of energy transferred as heat. Such a definition is customarily based on the work of Carathéodory (1909), referring to processes in a closed system, as follows.

The internal energy U_X of a body in an arbitrary state *X* can be determined by amounts of work adiabatically performed by the body on its surrounds when it starts from a reference state *O*. Such work is assessed through quantities defined in the surroundings of the body. It is supposed that such work can be assessed accurately, without error due to friction in the surroundings; friction in the body is not excluded by this definition. The adiabatic performance of work is defined in terms of adiabatic walls, which allow transfer of energy as work, but no other transfer, of energy or matter. In particular they do not allow the passage of energy as heat. According to this definition, work performed adiabatically is in general accompanied by friction within the thermodynamic system or body. On the other hand, according to Carathéodory (1909), there also exist non-adiabatic walls, which are postulated to be "permeable only to heat", and are called diathermal.

For the definition of quantity of energy transferred as heat, it is customarily envisaged that an arbitrary state of interest *Y* is reached from state *O* by a process with two components, one adiabatic and the other not adiabatic. For convenience one may say that the adiabatic component was the sum of work done by the body through volume change through movement of the walls while the non-adiabatic wall was temporarily rendered adiabatic, and of isochoric adiabatic work. Then the non-adiabatic component is a process of energy transfer through the wall that passes only heat, newly made accessible for the purpose of this transfer, from the surroundings to the body. The change in internal energy to reach the state Y from the state O is the difference of the two amounts of energy transferred.

Although Carathéodory himself did not state such a definition, following his work it is customary in theoretical studies to define the quantity of energy transferred as heat, Q, to the body from its surroundings, in the combined process of change to state Y from the state O, as the change in internal energy, ΔU_Y, minus the amount of work, W, done by the body on its surrounds by the adiabatic process, so that $Q = \Delta U_Y - W$.

In this definition, for the sake of conceptual rigour, the quantity of energy transferred as heat is not specified directly in terms of the non-adiabatic process. It is defined through knowledge of precisely two variables, the change of internal energy and the amount of adiabatic work done, for the combined process of change from the reference state O to the arbitrary state Y. It is important that this does not explicitly involve the amount of energy transferred in the non-adiabatic component of the combined process. It is assumed here that the amount of energy required to pass from state O to state Y, the change of internal energy, is known, independently of the combined process, by a determination through a purely adiabatic process, like that for the determination of the internal energy of state X above. The rigour that is prized in this definition is that there is one and only one kind of energy transfer admitted as fundamental: energy transferred as work. Energy transfer as heat is considered as a derived quantity. The uniqueness of work in this scheme is considered to guarantee rigor and purity of conception. The conceptual purity of this definition, based on the concept of energy transferred as work as an ideal notion, relies on the idea that some frictionless and otherwise non-dissipative processes of energy transfer can be realized in physical actuality. The second law of thermodynamics, on the other hand, assures us that such processes are not found in nature.

Heat, Temperature, and Thermal Equilibrium Regarded as Jointly Primitive Notions

Before the rigorous mathematical definition of heat based on Carathéodory's 1909 paper, recounted just above, historically, heat, temperature, and thermal equilibrium were presented in thermodynamics textbooks as jointly primitive notions. Carathéodory introduced his 1909 paper thus: "The proposition that the discipline of thermodynamics can be justified without recourse to any hypothesis that cannot be verified experimentally must be regarded as one of the most noteworthy results of the research in thermodynamics that was accomplished during the last century." Referring to the "point of view adopted by most authors who were active in the last fifty years", Carathéodory wrote: "There exists a physical quantity called heat that is not identical with the mechanical quantities (mass, force, pressure, etc.) and whose variations can be determined by calorimetric measurements." James Serrin introduces an account of the theory of thermodynamics thus: "In the following section, we shall use the classical notions of *heat*, *work*, and *hotness* as primitive elements, ... That heat is an appropriate and natural primitive for thermodynamics was already accepted by Carnot. Its continued validity as a primitive element of thermodynamical structure is due to the fact that it synthesizes an essential physical concept, as well as to its successful use in recent work to unify different constitutive theories." This traditional kind of presentation of the basis of thermodynamics includes ideas that may be summarized by the statement that heat transfer is purely due to spatial non-uniformity of temperature, and is by conduction and radiation, from hotter to colder bodies. It is sometimes proposed that this traditional kind of presentation necessarily rests on "circular reasoning"; against this proposal, there stands the rigorously logical mathematical development of the theory presented by Truesdell and Bharatha (1977).

This alternative approach to the definition of quantity of energy transferred as heat differs in logical structure from that of Carathéodory, recounted just above.

This alternative approach admits calorimetry as a primary or direct way to measure quantity of energy transferred as heat. It relies on temperature as one of its primitive concepts, and used in calorimetry. It is presupposed that enough processes exist physically to allow measurement of differences in internal energies. Such processes are not restricted to adiabatic transfers of energy as work. They include calorimetry, which is the commonest practical way of finding internal energy differences. The needed temperature can be either empirical or absolute thermodynamic.

In contrast, the Carathéodory way recounted just above does not use calorimetry or temperature in its primary definition of quantity of energy transferred as heat. The Carathéodory way regards calorimetry only as a secondary or indirect way of measuring quantity of energy transferred as heat. As recounted in more detail just above, the Carathéodory way regards quantity of energy transferred as heat in a process as primarily or directly defined as a residual quantity. It is calculated from the difference of the internal energies of the initial and final states of the system, and from the actual work done by the system during the process. That internal energy difference is supposed to have been measured in advance through processes of purely adiabatic transfer of energy as work, processes that take the system between the initial and final states. By the Carathéodory way it is presupposed as known from experiment that there actually physically exist enough such adiabatic processes, so that there need be no recourse to calorimetry for measurement of quantity of energy transferred as heat. This presupposition is essential but is explicitly labeled neither as a law of thermodynamics nor as an axiom of the Carathéodory way. In fact, the actual physical existence of such adiabatic processes is indeed mostly supposition, and those supposed processes have in most cases not been actually verified empirically to exist.

Heat Transfer in Engineering

The discipline of heat transfer, typically considered an aspect of mechanical engineering and chemical engineering, deals with specific applied methods by which thermal energy in a system is generated, or converted, or transferred to another system. Although the definition of heat implicitly means the transfer of energy, the term *heat transfer* encompasses this traditional usage in many engineering disciplines and laymen language.

Heat transfer includes the mechanisms of heat conduction, thermal radiation, and mass transfer.

In engineering, the term *convective heat transfer* is used to describe the combined effects of conduction and fluid flow. From the thermodynamic point of view, heat flows into a fluid by diffusion to increase its energy, the fluid then transfers (advects) this increased internal energy (not heat) from one location to another, and this is then followed by a second thermal interaction which transfers heat to a second body or system, again by diffusion. This entire process is often regarded as an additional mechanism of heat transfer, although technically, "heat transfer" and thus heating and cooling occurs only on either end of such a conductive flow, but not as a result of flow. Thus, conduction can be said to "transfer" heat only as a net result of the process, but may not do so at every time within the complicated convective process.

Although distinct physical laws may describe the behavior of each of these methods, real systems often exhibit a complicated combination which are often described by a variety of mathematical methods.

3

Modes of Heat Transfer

The transferring of heat by microscopic collisions within a body is known as thermal conduction. Conduction as a process takes place in every phase of matter, solids, liquids, gases and plasma. The alternative modes of heat transfer explained in this section are convective heat transfer and radiation. The topics discussed in the chapter are of great importance to broaden the existing knowledge on heat transfer.

Thermal Conduction

Thermal conduction is the transfer of heat (internal energy) by microscopic collisions of particles and movement of electrons within a body. The microscopically colliding objects, that include molecules, atoms, and electrons, transfer disorganized microscopic kinetic and potential energy, jointly known as internal energy. Conduction takes place in all phases of matter, such as solids, liquids, gases and plasmas. The rate at which energy is conducted as heat between two bodies is a function of the temperature difference (temperature gradient) between the two bodies and the properties of the conductive medium through which the heat is transferred. Thermal conduction was originally called diffusion.

Heat spontaneously flows from a hotter to a colder body. For example, heat is conducted from the hotplate of an electric stove to the bottom of a saucepan in contact with it. In the absence of an external driving energy source to the contrary, within a body or between bodies, temperature differences decay over time, and thermal equilibrium is approached, temperature becoming more uniform.

In conduction, the heat flow is within and through the body itself. In contrast, in heat transfer by thermal radiation, the transfer is often between bodies, which may be separated spatially. Also possible is transfer of heat by a combination of conduction and thermal radiation. In convection, internal energy is carried between bodies by a moving material carrier. In solids, conduction is mediated by the combination of vibrations and collisions of molecules, of propagation and collisions of phonons, and of diffusion and collisions of free electrons. In gases and liquids, conduction is due to the collisions and diffusion of molecules during their random motion. Photons in this context do not collide with one another, and so heat transport by electromagnetic radiation is conceptually distinct from heat conduction by microscopic diffusion and collisions of material particles and phonons. But the distinction is often not easily observed, unless the material is semi-transparent.

In the engineering sciences, heat transfer includes the processes of thermal radiation, convection, and sometimes mass transfer. Usually, more than one of these processes occurs in a given situation. The conventional symbol for the material property, thermal conductivity, is .

Overview

On a microscopic scale, conduction occurs within a body considered as being stationary; this means that the kinetic and potential energies of the bulk motion of the body are separately accounted for. Internal energy diffuses as rapidly moving or vibrating atoms and molecules interact with neighboring particles, transferring some of their microscopic kinetic and potential energies, these quantities being defined relative to the bulk of the body considered as being stationary. Heat is transferred by conduction when adjacent atoms or molecules collide, or as several electrons move backwards and forwards from atom to atom in a disorganized way so as not to form a macroscopic electric current, or as phonons collide and scatter. Conduction is the most significant means of heat transfer within a solid or between solid objects in thermal contact. Conduction is greater in solids because the network of relatively close fixed spatial relationships between atoms helps to transfer energy between them by vibration.

Fluids (and especially gases) are less conductive. This is due to the large distance between atoms in a gas: fewer collisions between atoms means less conduction. The conductivity of gases increases with temperature. Conductivity increases with increasing pressure from vacuum up to a critical point that the density of the gas is such that molecules of the gas may be expected to collide with each other before they transfer heat from one surface to another. After this point, conductivity increases only slightly with increasing pressure and density.

Thermal contact conductance is the study of heat conduction between solid bodies in contact. A temperature drop is often observed at the interface between the two surfaces in contact. This phenomenon is said to be a result of a thermal contact resistance existing between the contacting surfaces. Interfacial thermal resistance is a measure of an interface's resistance to thermal flow. This thermal resistance differs from contact resistance, as it exists even at atomically perfect interfaces. Understanding the thermal resistance at the interface between two materials is of primary significance in the study of its thermal properties. Interfaces often contribute significantly to the observed properties of the materials.

The inter-molecular transfer of energy could be primarily by elastic impact, as in fluids, or by free electron diffusion, as in metals, or phonon vibration, as in insulators. In insulators, the heat flux is carried almost entirely by phonon vibrations.

Metals (e.g., copper, platinum, gold, etc.) are usually good conductors of thermal energy.

This is due to the way that metals bond chemically: metallic bonds (as opposed to covalent or ionic bonds) have free-moving electrons that transfer thermal energy rapidly through the metal. The *electron fluid* of a conductive metallic solid conducts most of the heat flux through the solid. Phonon flux is still present, but carries less of the energy. Electrons also conduct electric current through conductive solids, and the thermal and electrical conductivities of most metals have about the same ratio. A good electrical conductor, such as copper, also conducts heat well. Thermoelectricity is caused by the interaction of heat flux and electric current. Heat conduction within a solid is directly analogous to diffusion of particles within a fluid, in the situation where there are no fluid currents.

To quantify the ease with which a particular medium conducts, engineers employ the thermal conductivity, also known as the conductivity constant or conduction coefficient, k. In thermal conductivity, k is defined as "the quantity of heat, Q, transmitted in time (t) through a thickness (L), in a direction normal to a surface of area (A), due to a temperature difference (ΔT) [...]". Thermal conductivity is a material *property* that is primarily dependent on the medium's phase, temperature, density, and molecular bonding. Thermal effusivity is a quantity derived from conductivity, which is a measure of its ability to exchange thermal energy with its surroundings.

Steady-state Conduction

Steady state conduction is the form of conduction that happens when the temperature difference(s) driving the conduction are constant, so that (after an equilibration time), the spatial distribution of temperatures (temperature field) in the conducting object does not change any further. Thus, all partial derivatives of temperature *with respect to space* may either be zero or have nonzero values, but all derivatives of temperature at any point *with respect to time* are uniformly zero. In steady state conduction, the amount of heat entering any region of an object is equal to amount of heat coming out (if this were not so, the temperature would be rising or falling, as thermal energy was tapped or trapped in a region).

For example, a bar may be cold at one end and hot at the other, but after a state of steady state conduction is reached, the spatial gradient of temperatures along the bar does not change any further, as time proceeds. Instead, the temperature at any given section of the rod remains constant, and this temperature varies linearly in space, along the direction of heat transfer.

In steady state conduction, all the laws of direct current electrical conduction can be applied to "heat currents". In such cases, it is possible to take "thermal resistances" as the analog to electrical resistances. In such cases, temperature plays the role of voltage, and heat transferred per unit time (heat power) is the analog of electric current. Steady state systems can be modelled by networks of such thermal resistances in series and in parallel, in exact analogy to electrical networks of resistors.

Transient Conduction

In general, during any period in which temperatures change *in time* at any place within an object, the mode of thermal energy flow is termed *transient conduction*. Another term is "non steady-state" conduction, referring to time-dependence of temperature fields in an object. Non-steady-state situations appear after an imposed change in temperature at a boundary of an object. They may also occur with temperature changes inside an object, as a result of a new source or sink of heat suddenly introduced within an object, causing temperatures near the source or sink to change in time.

When a new perturbation of temperature of this type happens, temperatures within the system change in time toward a new equilibrium with the new conditions, provided that these do not change. After equilibrium, heat flow into the system once again equals the heat flow out, and temperatures at each point inside the system no longer change. Once this happens, transient conduction is ended, although steady-state conduction may continue if heat flow continues.

If changes in external temperatures or internal heat generation changes are too rapid for the equilibrium of temperatures in space to take place, then the system never reaches a state of unchanging temperature distribution in time, and the system remains in a transient state.

An example of a new source of heat "turning on" within an object, causing transient conduction, is an engine starting in an automobile. In this case, the transient thermal conduction phase for the entire machine is over, and the steady state phase appears, as soon as the engine reaches steady-state operating temperature. In this state of steady-state equilibrium, temperatures vary greatly from the engine cylinders to other parts of the automobile, but at no point in space within the automobile does temperature increase or decrease. After establishing this state, the transient conduction phase of heat transfer is over.

New external conditions also cause this process: for example the copper bar in the example steady-state conduction experiences transient conduction as soon as one end is subjected to a different temperature from the other. Over time, the field of temperatures inside the bar reach a new steady-state, in which a constant temperature gradient along the bar is finally set up, and this gradient then stays constant in space. Typically, such a new steady state gradient is approached exponentially with time after a new temperature-or-heat source or sink, has been introduced. When a "transient conduction" phase is over, heat flow may still continue at high power, so long as temperatures do not change.

An example of transient conduction that does not end with steady-state conduction, but rather no conduction, occurs when a hot copper ball is dropped into oil at a low temperature. Here, the temperature field within the object begins to change as a function of time, as the heat is removed from the metal, and the interest lies in analyzing this spatial change of temperature within the object over time, until all gradients disappear entirely (the ball has reached the same temperature as the oil). Mathematically, this condition is also approached exponentially; in theory it takes infinite time, but in practice it is over, for all intents and purposes, in a much shorter period. At the end of this process with no heat sink but the internal parts of the ball (which are finite), there is no steady state heat conduction to reach. Such a state never occurs in this situation, but rather the end of the process is when there is no heat conduction at all.

The analysis of non steady-state conduction systems is more complex than that of steady-state systems. If the conducting body has a simple shape, then exact analytical mathematical expressions and solutions may be possible. However, most often, because of complicated shapes with varying thermal conductivities within the shape (i.e., most complex objects, mechanisms or machines in engineering) often the application of approximate theories is required, and/or numerical analysis by computer. One popular graphical method involves the use of Heisler Charts.

Occasionally, transient conduction problems may be considerably simplified if regions of the object being heated or cooled can be identified, for which thermal conductivity is very much greater than that for heat paths leading into the region. In this case, the region with high conductivity can often be treated in the lumped capacitance model, as a "lump" of material with a simple thermal capacitance consisting of its aggregate heat capacity. Such regions warm or cool, but show no significant temperature *variation* across their extent, during the process (as compared to the rest of the system). This is due to their far higher conductance. During transient conduction, therefore, the temperature across their conductive regions changes uniformly in space, and as a simple exponential in time. An example of such systems are those that follow Newton's law of cooling during transient cooling (or the reverse during heating). The equivalent thermal circuit consists of a simple capacitor in series with a resistor. In such cases, the remainder of the system with high thermal resistance (comparatively low conductivity) plays the role of the resistor in the circuit.

Relativistic Conduction

The theory of relativistic heat conduction is a model that is compatible with the theory of special relativity. For most of the last century, it was recognized that the Fourier equation is in contradiction with the theory of relativity because it admits an infinite speed of propagation of heat signals. For example, according to the Fourier equation, a pulse of heat at the origin would be felt at infinity instantaneously. The speed of information propagation is faster than the speed of light in vacuum, which is physically inadmissible within the framework of relativity.

Quantum Conduction

Second sound is a quantum mechanical phenomenon in which heat transfer occurs by wave-like motion, rather than by the more usual mechanism of diffusion. Heat takes the place of pressure in normal sound waves. This leads to a very high thermal conductivity. It is known as "second sound" because the wave motion of heat is similar to the propagation of sound in air.

Fourier's Law

The law of heat conduction, also known as Fourier's law, states that the time rate of heat transfer through a material is proportional to the negative gradient in the temperature and to the area, at right angles to that gradient, through which the heat flows. We can state this law in two equivalent forms: the integral form, in which we look at the amount of energy flowing into or out of a body as a whole, and the differential form, in which we look at the flow rates or fluxes of energy locally.

Newton's law of cooling is a discrete analog of Fourier's law, while Ohm's law is the electrical analogue of Fourier's law.

Differential Form

The differential form of Fourier's law of thermal conduction shows that the local heat flux density, $\vec{q}$, is equal to the product of thermal conductivity, k, and the negative local temperature gradient, $-\nabla T$. The heat flux density is the amount of energy that flows through a unit area per unit time.

$$\vec{q} = -k\nabla T$$

where (including the SI units)

$\vec{q}$ is the local heat flux density, $W \cdot m^{-2}$

k is the material's conductivity, $W \cdot m^{-1} \cdot K^{-1}$,

∇T is the temperature gradient, $K \cdot m^{-1}$.

The thermal conductivity, k, is often treated as a constant, though this is not always true. While the thermal conductivity of a material generally varies with temperature, the variation can be small over a significant range of temperatures for some common materials. In anisotropic materials, the thermal conductivity typically varies with orientation; in this case k is represented by a second-order tensor. In non-uniform materials, k varies with spatial location.

For many simple applications, Fourier's law is used in its one-dimensional form. In the x-direction,

$$q_x = -k\frac{dT}{dx}$$

Integral Form

By integrating the differential form over the material's total surface , we arrive at the integral form of Fourier's law:

$$\frac{\partial Q}{\partial t} = -k \oiint \nabla T \cdot dS$$

where (including the SI units):

- $\frac{\partial Q}{\partial t}$ is the amount of heat transferred per unit time (in W), and
- dS is an oriented surface area element (in m²)

The above differential equation, when integrated for a homogeneous material of 1-D geometry between two endpoints at constant temperature, gives the heat flow rate as:

$$\frac{\Delta Q}{\Delta t} = -kA\frac{\Delta T}{\Delta x}$$

where

A is the cross-sectional surface area,

ΔT is the temperature difference between the ends,

Δx is the distance between the ends.

This law forms the basis for the derivation of the heat equation.

Conductance

Writing

$$U = \frac{k}{\Delta x},$$

where U is the conductance, in W/(m² K).

Fourier's law can also be stated as:

$$\frac{\Delta Q}{\Delta t} = UA(-\Delta T).$$

The reciprocal of conductance is resistance, R, given by:

$$R = \frac{1}{U} = \frac{\Delta x}{k} = \frac{A(-\Delta T)}{\frac{\Delta Q}{\Delta t}}.$$

Resistance is additive when several conducting layers lie between the hot and cool regions, because A and Q are the same for all layers. In a multilayer partition, the total conductance is related to the conductance of its layers by:

$$\frac{1}{U}=\frac{1}{U_1}+\frac{1}{U_2}+\frac{1}{U_3}+\cdots$$

So, when dealing with a multilayer partition, the following formula is usually used:

$$\frac{\Delta Q}{\Delta t}=\frac{A(-\Delta T)}{\frac{\Delta x_1}{k_1}+\frac{\Delta x_2}{k_2}+\frac{\Delta x_3}{k_3}+\cdots}.$$

For heat conduction from one fluid to another through a barrier, it is sometimes important to consider the conductance of the thin film of fluid that remains stationary next to the barrier. This thin film of fluid is difficult to quantify because its characteristics depend upon complex conditions of turbulence and viscosity—but when dealing with thin high-conductance barriers it can sometimes be quite significant.

Intensive-property Representation

The previous conductance equations, written in terms of extensive properties, can be reformulated in terms of intensive properties. Ideally, the formulae for conductance should produce a quantity with dimensions independent of distance, like Ohm's Law for electrical resistance, $R=V/I$, and conductance, $G=I/V$.

From the electrical formula: $R=\rho x/A$, where ρ is resistivity, x is length, and A is cross-sectional area, we have , $G=kA/x$ where G is conductance, k is conductivity, x is length, and A is cross-sectional area.

For Heat,

$$U=\frac{kA}{\ddot{A}x},$$

where U is the conductance.

Fourier's law can also be stated as:

$$\dot{Q}=U\Delta T,$$

analogous to Ohm's law, $I=V/R$ or $I=VG$.

The reciprocal of conductance is resistance, R, given by:

$$R = \frac{\Delta T}{\dot{Q}},$$

analogous to Ohm's law, $R = V / I$.

The rules for combining resistances and conductances (in series and in parallel) are the same for both heat flow and electric current.

Cylindrical Shells

Conduction through cylindrical shells (e.g. pipes) can be calculated from the internal radius, r_1, the external radius, r_2, the length, ℓ, and the temperature difference between the inner and outer wall, $T_2 - T_1$.

The surface area of the cylinder is $A_r = 2\pi r\ell$

When Fourier's equation is applied:

$$\dot{Q} = -kA_r \frac{\mathrm{d}T}{\mathrm{d}r} = -2k\pi r\ell \frac{\mathrm{d}T}{\mathrm{d}r}$$

and rearranged:

$$\dot{Q}\int_{r_1}^{r_2} \frac{1}{r}\mathrm{d}r = -2k\pi\ell\int_{T_1}^{T_2} \mathrm{d}T$$

then the rate of heat transfer is:

$$\dot{Q} = 2k\pi\ell \frac{T_1 - T_2}{\ln(r_2 / r_1)}$$

the thermal resistance is:

$$R_c = \frac{\Delta T}{\dot{Q}} = \frac{\ln(r_2 / r_1)}{2\pi k\ell}$$

and $\dot{Q} = 2\pi k\ell r_m \frac{T_1 - T_2}{r_2 - r_1}$, where $r_m = \frac{r_2 - r_1}{\ln(r_2 / r_1)}$. It is important to note that this is the log-mean radius.

Spherical

The conduction through a spherical shell with internal radius, r_1, and external radius, r_2, can be calculated in a similar manner as for a cylindrical shell.

The surface area of the sphere is: $A = 4\pi r^2$.

Solving in a similar manner as for a cylindrical shell produces:

$$\dot{Q} = 4k\pi \frac{T_1 - T_2}{1/r_1 - 1/r_2} = 4k\pi \frac{(T_1 - T_2)r_1 r_2}{r_2 - r_1}$$

Transient Thermal Conduction

Interface Heat Transfer

The heat transfer at an interface is considered a transient heat flow. To analyze this problem, the Biot number is important to understand how the system behaves. The Biot number is determined by: $Bi = \frac{hL}{k}$

The heat transfer coefficient h , is introduced in this formula, and is measured in $\frac{J}{m^2 sK}$.

If the system has a Biot number of less than 0.1, the material behaves according to New tonian cooling, i.e. with negligible temperature gradient within the body. If the Biot number is greater than 0.1, the system behaves as a series solution. The temperature profile in terms of time can be determined by the function can be derived from the equation

$$q = -h\Delta T,$$

which becomes

$$\frac{T - T_f}{T_i - T_f} = \exp\left[\frac{-hAt}{\rho C_p V}\right].$$

Theheattransfercoefficient,h,ismeasuredin $\frac{\text{W}}{\text{m}^2\text{K}}$, ,andrepresentsthetransferofheatatan interface between two materials. This value is different at every interface, and is an important concept in understanding heat flow at an interface.

The series solution can be analyzed with a nomogram. A nomogram has relative temperature as the y coordinate and the Fourier number, which is calculated by

$$Fo = \frac{\alpha t}{L^2}.$$

The Biot number increases as the Fourier number decreases. There are five steps to determine a temperature profile in terms of time.

1. Calculate the Biot number
2. Determine which relative depth matters, either x or L.

3. Convert time to the Fourier number.
4. Convert T_i to relative temperature with the boundary conditions.
5. Compared required point to trace specified Biot number on the nomogram.

Thermal Conduction Applications

Splat Cooling

Splat cooling is a method for quenching small droplets of molten materials by rapid contact with a cold surface. The particles undergo a characteristic cooling process, with the heat profile at $t = 0$ for initial temperature as the maximum at $x = 0$ and $T = 0$ at $x = -\infty$ and $x = \infty$, and the heat profile at $t = \infty$ for $-\infty \leq x \leq \infty$ as the boundary conditions. Splat cooling rapidly ends in a steady state temperature, and is similar in form to the Gaussian diffusion equation. The temperature profile, with respect to the position and time of this type of cooling, varies with:

$$T(x,t) - T_i = \frac{T_i \text{ÄX}}{2\sqrt{\pi \alpha t}} \exp\left(-\frac{x^2}{4\alpha t}\right)$$

Splat cooling is a fundamental concept that has been adapted for practical use in the form of thermal spraying. The thermal diffusivity coefficient, represented as α, can be written as $\alpha = \frac{k}{\rho C_p}$. This varies according to the material.

Metal Quenching

Metal quenching is a transient heat transfer process in terms of the Time Temperature Transformation (TTT). It is possible to manipulate the cooling process to adjust the phase of a suitable material. For example, appropriate quenching of steel can convert a desirable proportion of its content of austenite to martensite, creating a very tough product. To achieve this, it is necessary to quench at the "nose" (or eutectic) of the TTT diagram. Since materials differ in their Biot numbers, the time it takes for the material to quench, or the Fourier number, varies in practice. In steel, the quenching temperature range is generally from 600 °C to 200 °C. To control the quenching time and to select suitable quenching media, it is necessary to determine the Fourier number from the desired quenching time, the relative temperature drop, and the relevant Biot number. Usually, the correct figures are read from a standard nomogram.By calculating the heat transfer coefficient from this Biot number, one can find a liquid medium suitable for the application.

Zeroth Law of Thermodynamics

One statement of the so-called zeroth law of thermodynamics is directly focused on the idea of conduction of heat. Bailyn (1994) writes that "... the zeroth law may be stated:

All diathermal walls are equivalent."

A diathermal wall is a physical connection between two bodies that allows the passage of heat between them. Bailyn is referring to diathermal walls that exclusively connect two bodies, especially conductive walls.

This statement of the 'zeroth law' belongs to an idealized theoretical discourse, and actual physical walls may have peculiarities that do not conform to its generality.

For example, the material of the wall must not undergo a phase transition, such as evaporation or fusion, at the temperature at which it must conduct heat. But when only thermal equilibrium is considered and time is not urgent, so that the conductivity of the material does not matter too much, one suitable heat conductor is as good as another. Conversely, another aspect of the zeroth law is that, subject again to suitable restrictions, a given diathermal wall is indifferent to the nature of the heat bath to which it is connected. For example, the glass bulb of a thermometer acts as a diathermal wall whether exposed to a gas or to a liquid, provided they do not corrode or melt it.

These indifferences are amongst the defining characteristics of heat transfer. In a sense, they are symmetries of heat transfer.

Convective Heat Transfer

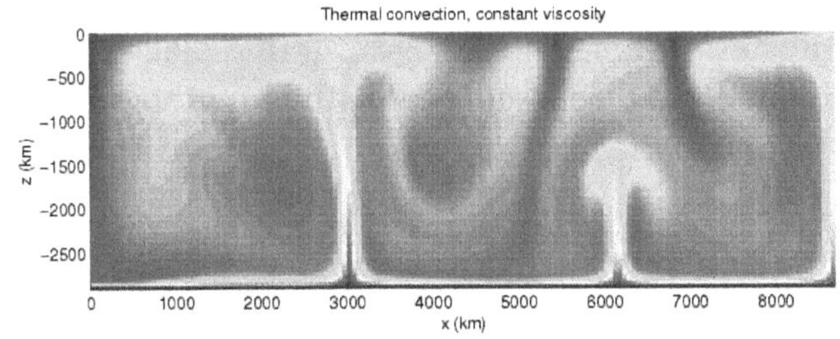

Simulation of thermal convection. Red hues designate hot areas, while regions with blue hues are cold. A hot, less-dense lower boundary layer sends plumes of hot material upwards, and likewise, cold material from the top moves downwards. This illustration is taken from a model of convection in the Earth's mantle.

Convective heat transfer, often referred to simply as convection, is the transfer of heat from one place to another by the movement of fluids. Convection is usually the dominant form of heat transfer(convection) in liquids and gases. Although often discussed as a distinct method of heat transfer, convective heat transfer involves the combined processes of conduction (heat diffusion) and advection (heat transfer by bulk fluid flow).

Convection can be "forced" by movement of a fluid by means other than buoyancy forces (for example, a water pump in an automobile engine). Thermal expansion of fluids may also force convection. In other cases, natural buoyancy forces alone are entirely responsible for fluid motion when the fluid is heated, and this process is called "natural convection". An example is the draft in a chimney or around any fire. In natural convection, an increase in temperature produces a reduction in density, which in turn causes fluid motion due to pressures and forces when fluids of different densities are affected by gravity (or any g-force). For example, when water is heated on a stove, hot water from the bottom of the pan rises, displacing the colder denser liquid, which falls. After heating has stopped, mixing and conduction from this natural convection eventually result in a nearly homogeneous density, and even temperature. Without the presence of gravity (or conditions that cause a g-force of any type), natural convection does not occur, and only forced-convection modes operate. t

The convection heat transfer mode comprises one mechanism. In addition to energy transfer due to specific molecular motion (diffusion), energy is transferred by bulk, or macroscopic, motion of the fluid. This motion is associated with the fact that, at any instant, large numbers of molecules are moving collectively or as aggregates. Such motion, in the presence of a temperature gradient, contributes to heat transfer. Because the molecules in aggregate retain their random motion, the total heat transfer is then due to the superposition of energy transport by random motion of the molecules and by the bulk motion of the fluid. It is customary to use the term convection when referring to this cumulative transport and the term advection when referring to the transport due to bulk fluid motion.

Overview

Convection is the transfer of thermal energy from one place to another by the movement of fluids. Although often discussed as a distinct method of heat transfer, convection describes the combined effects of conduction and fluid flow or mass exchange.

Two types of convective heat transfer may be distinguished:

- Free or natural convection: when fluid motion is caused by buoyancy forces that result from the density variations due to variations of thermal temperature in the fluid. In the absence of an external source, when the fluid is in contact with a hot surface, its molecules separate and scatter, causing the fluid to be less dense. As a consequence, the fluid is displaced while the cooler fluid gets denser and the fluid sinks. Thus, the hotter volume transfers heat towards the cooler volume of that fluid. Familiar examples are the upward flow of air due to a fire or hot object and the circulation of water in a pot that is heated from below.
- Forced convection: when a fluid is forced to flow over the surface by an external source such as fans, by stirring, and pumps, creating an artificially induced convection current.

Internal and external flow can also classify convection. Internal flow occurs when a fluid is enclosed by a solid boundary such when flowing through a pipe. An external flow occurs when a fluid extends indefinitely without encountering a solid surface. Both of these types of convection, either natural or forced, can be internal or external because they are independent of each other.The bulk temperature, or the average fluid temperature, is a convenient reference point for evaluating properties related to convective heat transfer, particularly in applications related to flow in pipes and ducts.

This color schlieren image reveals thermal convection from a human hand (in silhouette form) to the surrounding still atmosphere. Photographed using schlieren equipment.

Papers lifted on rising convective air current from warm radiator

Further classification can be made depending on the smoothness and undulations of the solid surfaces. Not all surfaces are smooth, though a bulk of the available information deals with smooth surfaces. Wavy irregular surfaces are commonly encountered in heat transfer devices which include solar collectors, regenerative heat exchangers and underground energy storage systems. They have a significant role to play in the heat transfer processes in these applications. Since they bring in an added complexity due to the undulations in the surfaces, they need to be tackled with mathematical finesse through elegant simplification techniques. Also they do affect the flow and heat transfer characteristics, thereby behaving differently from straight smooth surfaces.

For a visual experience of natural convection, a glass filled with hot water and some red food dye may be placed inside a fish tank with cold, clear water. The convection currents of the red liquid may be seen to rise and fall in different regions, then eventually settle, illustrating the process as heat gradients are dissipated.

Newton's Law of Cooling

Convection-cooling is sometimes loosely assumed to be described by Newton's law of cooling.

Newton's law states that *the rate of heat loss of a body is proportional to the difference in temperatures between the body and its surroundings while under the effects of a breeze*. The constant of proportionality is the heat transfer coefficient. The law applies when the coefficient is independent, or relatively independent, of the temperature difference between object and environment.

In classical natural convective heat transfer, the heat transfer coefficient is dependent on the temperature. However, Newton's law does approximate reality when the temperature changes are relatively small.

Convective Heat Transfer

The basic relationship for heat transfer by convection is:

$$\dot{Q} = hA(T_a - T_b)$$

where $\dot{Q}$ is the heat transferred per unit time, A is the area of the object, h is the heat transfer coefficient, T_a is the object's surface temperature and T_b is the fluid temperature.

The convective heat transfer coefficient is dependent upon the physical properties of the fluid and the physical situation. Values of h have been measured and tabulated for commonly encountered fluids and flow of situations.

Radiation

In physics, radiation is the emission or transmission of energy in the form of waves or particles through space or through a material medium. This includes:

- electromagnetic radiation, such as radio waves, visible light, x-rays, and gamma radiation (γ)
- particle radiation, such as alpha radiation (α), beta radiation (β), and neutron radiation (particles of non-zero rest energy)
- acoustic radiation, such as ultrasound, sound, and seismic waves (dependent on a physical transmission medium)

- gravitational radiation, radiation that takes the form of gravitational waves, or ripples in the curvature of spacetime.

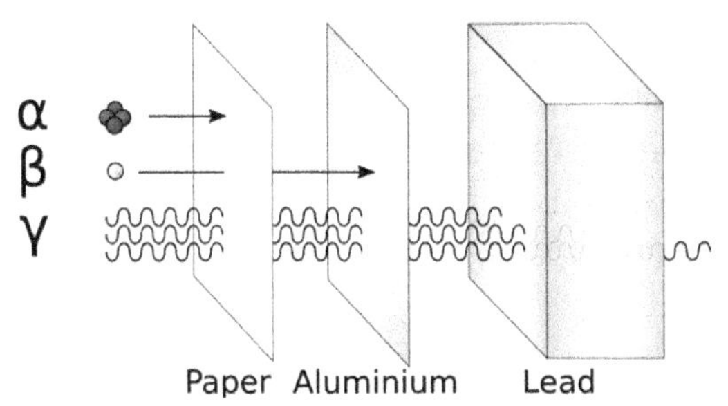

Illustration of the relative abilities of three different types of ionizing radiation to penetrate solid matter. Typical alpha particles (α) are stopped by a sheet of paper, while beta particles (β) are stopped by an aluminum plate. Gamma radiation (γ) is damped when it penetrates lead. Note caveats in the text about this simplified diagram.

The international symbol for types and levels of radiation that are unsafe for unshielded humans. Radiation in general exists throughout nature, such as in light and sound.

Radiation is often categorized as either ionizing or non-ionizing depending on the energy of the radiated particles. Ionizing radiation carries more than 10 eV, which is enough to ionize atoms and molecules, and break chemical bonds. This is an important distinction due to the large difference in harmfulness to living organisms. A common source of ionizing radiation is radioactive materials that emit α, β, or γ radiation, consisting of helium nuclei, electrons or positrons, and photons, respectively. Other sources include X-rays from medical radiography examinations and muons, mesons, positrons, neutrons and other particles that constitute the secondary cosmic rays that are produced after primary cosmic rays interact with Earth's atmosphere.

Gamma rays, X-rays and the higher energy range of ultraviolet light constitute the ionizing part of the electromagnetic spectrum. The lower-energy, longer-wavelength part of the spectrum including visible light, infrared light, microwaves, and radio waves is non-ionizing; its main effect when interacting with tissue is heating. This type of radiation only damages cells if the intensity is high enough to cause excessive heating. Ultraviolet radiation has some features of both ionizing and non-ionizing radiation. While the part of the ultraviolet spectrum that penetrates the Earth's atmosphere is non-ionizing, this radiation does far more damage to many molecules in biological systems than can be accounted for by heating effects, sunburn being a well-known example. These properties derive from ultraviolet's power to alter chemical bonds, even without having quite enough energy to ionize atoms.

The word radiation arises from the phenomenon of waves *radiating* (i.e., traveling outward in all directions) from a source. This aspect leads to a system of measurements and physical units that are applicable to all types of radiation. Because such radiation expands as it passes through space, and as its energy is conserved (in vacuum), the intensity of all types of radiation from a point source follows an inverse-square law in relation to the distance from its source. This law does not apply close to an extended source of radiation or for focused beams.

Ionizing Radiation

Radiation with sufficiently high energy can ionize atoms; that is to say it can knock electrons off atoms and create ions. Ionization occurs when an electron is stripped (or "knocked out") from an electron shell of the atom, which leaves the atom with a net positive charge. Because living cells and, more importantly, the DNA in those cells can be damaged by this ionization, exposure to ionizing radiation is considered to increase the risk of cancer. Thus "ionizing radiation" is somewhat artificially separated from particle radiation and electromagnetic radiation, simply due to its great potential for biological damage. While an individual cell is made of trillions of atoms, only a small fraction of those will be ionized at low to moderate radiation powers. The probability of ionizing radiation causing cancer is dependent upon the absorbed dose of the radiation, and is a function of the damaging tendency of the type of radiation (equivalent dose) and the sensitivity of the irradiated organism or tissue (effective dose).

If the source of the ionizing radiation is a radioactive material or a nuclear process such as fission or fusion, there is particle radiation to consider. Particle radiation is subatomic particles accelerated to relativistic speeds by nuclear reactions. Because of their momenta they are quite capable of knocking out electrons and ionizing materials, but since most have an electrical charge, they don't have the penetrating power of ionizing radiation. The exception is neutron particles. There are several different kinds of these particles, but the majority are alpha particles, beta particles, neutrons, and protons. Roughly speaking, photons and particles with energies above about 10 electron volts (eV) are ionizing (some authorities use 33 eV, the ionization energy for water). Particle radiation from radioactive material or cosmic rays almost invariably carries enough energy to be ionizing.

Much ionizing radiation originates from radioactive materials and space (cosmic rays), and as such is naturally present in the environment, since most rock and soil has small concentrations of radioactive materials. The radiation is invisible and not directly detectable by human senses; as a result, instruments such as Geiger counters are usually required to detect its presence. In some cases, it may lead to secondary emission of visible light upon its interaction with matter, as in the case of Cherenkov radiation and radio-luminescence.

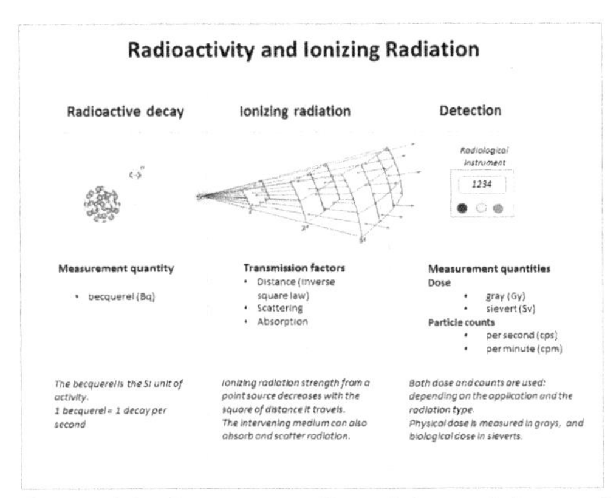

Graphic showing relationships between radioactivity and detected ionizing radiation

Ionizing radiation has many practical uses in medicine, research and construction, but presents a health hazard if used improperly. Exposure to radiation causes damage to living tissue; high doses result in Acute radiation syndrome (ARS), with skin burns, hair loss, internal organ failure and death, while any dose may result in an increased chance of cancer and genetic damage; a particular form of cancer, thyroid cancer, often occurs when nuclear weapons and reactors are the radiation source because of the biological proclivities of the radioactive iodine fission product, iodine-131. However, calculating the exact risk and chance of cancer forming in cells caused by ionizing radiation is still not well understood and currently estimates are loosely determined by population based on data from the atomic bombing in Japan and from reactor accident follow-up, such as with the Chernobyl disaster. The International Commission on Radiological Protection states that "The Commission is aware of uncertainties and lack of precision of the models and parameter values", "Collective effective dose is not intended as a tool for epidemiological risk assessment, and it is inappropriate to use it in risk projections" and "in particular, the calculation of the number of cancer deaths based on collective effectivc doses from trivial individual doses should be avoided."

Ultraviolet Radiation

Ultraviolet, of wavelengths from 10 nm to 125 nm, ionizes air molecules, causing it to be strongly absorbed by air and by ozone (O_3) in particular. Ionizing UV therefore does not penetrate Earth's atmosphere to a significant degree, and is sometimes referred to as vacuum ultraviolet. Although present in space, this part of the UV spectrum is not of biological importance, because it does not reach living organisms on Earth.

There is a zone of the atmosphere in which ozone absorbs some 98% of non-ionizing but dangerous UV-C and UV-B. This so-called ozone layer, starts at about 20 miles (32 km) and extends upward. Some of the ultraviolet spectrum that does reach the ground (the part that begins above energies of 3.1 eV, a wavelength less than 400 nm) is non-ionizing, but is still biologically hazardous due to the ability of single photons of this energy to cause electronic excitation in biological molecules, and thus damage them by means of unwanted reactions. An example is the formation of pyrimidine dimers in DNA, which begins at wavelengths below 365 nm (3.4 eV), which is well below ionization energy. This property gives the ultraviolet spectrum some of the dangers of ionizing radiation in biological systems without actual ionization occurring. In contrast, visible light and longer-wavelength electromagnetic radiation, such as infrared, microwaves, and radio waves, consists of photons with too little energy to cause damaging molecular excitation, and thus this radiation is far less hazardous per unit of energy.

X-ray

X-rays are electromagnetic waves with a wavelength less than about 10^{-9} m (greater than $3x10^{17}$ Hz and 1,240 eV). A smaller wavelength corresponds to a higher energy according to the equation $E=hc/\lambda$. ("E" is Energy; "h" is Planck's constant; "c" is the speed of light; "λ" is wavelength.) When an X-ray photon collides with an atom, the atom may absorb the energy of the photon and boost an electron to a higher orbital level or if the photon is very energetic, it may knock an electron from the atom altogether, causing the atom to ionize. Generally, larger atoms are more likely to absorb an X-ray photon since they have greater energy differences between orbital electrons. Soft tissue in the human body is composed of smaller atoms than the calcium atoms that make up bone, hence there is a contrast in the absorption of X-rays. X-ray machines are specifically designed to take advantage of the absorption difference between bone and soft tissue, allowing physicians to examine structure in the human body.

X-rays are also totally absorbed by the thickness of the earth's atmosphere, resulting in the prevention of the X-ray output of the sun, smaller in quantity than that of UV but nonetheless powerful, from reaching the surface.

Gamma Radiation

Gamma (γ) radiation consists of photons with a wavelength less than $3x10^{-11}$ meters (greater than 10^{19} Hz and 41.4 keV). Gamma radiation emission is a nuclear process that occurs to rid an unstable nucleus of excess energy after most nuclear reactions. Both alpha and beta particles have an electric charge and mass, and thus are quite likely to interact with other atoms in their path. Gamma radiation, however, is composed of photons, which have neither mass nor electric charge and, as a result, penetrates much further through matter than either alpha or beta radiation.

Gamma rays can be stopped by a sufficiently thick or dense layer of material, where the stopping power of the material per given area depends mostly (but not entirely) on the total mass along the path of the radiation, regardless of whether the material is of high or low density. However, as is the case with X-rays, materials with high atomic number such as lead or depleted uranium add a modest (typically 20% to 30%) amount of stopping power over an equal mass of less dense and lower atomic weight materials (such as water or concrete). The atmosphere absorbs all gamma rays approaching Earth from space. Even air is capable of absorbing gamma rays, halving the energy of such waves by passing through, on the average, 500 ft (150 m).

Alpha Radiation

Alpha particles are helium-4 nuclei (two protons and two neutrons). They interact with matter strongly due to their charges and combined mass, and at their usual velocities only penetrate a few centimeters of air, or a few millimeters of low density material (such as the thin mica material which is specially placed in some Geiger counter tubes to allow alpha particles in). This means that alpha particles from ordinary alpha decay do not penetrate the outer layers of dead skin cells and cause no damage to the live tissues below. Some very high energy alpha particles compose about 10% of cosmic rays, and these are capable of penetrating the body and even thin metal plates. However, they are of danger only to astronauts, since they are deflected by the Earth's magnetic field and then stopped by its atmosphere.

Alpha radiation is dangerous when alpha-emitting radioisotopes are ingested (breathed or swallowed). This brings the radioisotope close enough to sensitive live tissue for the alpha radiation to damage cells. Per unit of energy, alpha particles are at least 20 times more effective at cell-damage as gamma rays and X-rays. Examples of highly poisonous alpha-emitters are all isotopes of radium, radon, and polonium, due to the amount of decay that occur in these short half-life materials.

Beta Radiation

Beta-minus (β^-) radiation consists of an energetic electron. It is more penetrating than alpha radiation, but less than gamma. Beta radiation from radioactive decay can be stopped with a few centimeters of plastic or a few millimeters of metal. It occurs when a neutron decays into a proton in a nucleus, releasing the beta particle and an antineutrino. Beta radiation from linac accelerators is far more energetic and penetrating than natural beta radiation. It is sometimes used therapeutically in radiotherapy to treat superficial tumors.

Beta-plus (β^+) radiation is the emission of positrons, which are the antimatter form of electrons. When a positron slows to speeds similar to those of electrons in the material, the positron will annihilate an electron, releasing two gamma photons of 511 keV in the process. Those two gamma photons will be traveling in (approximately) opposite direction. The gamma radiation from positron annihilation consists of high energy photons, and is also ionizing.

Neutron Radiation

Neutrons are categorized according to their speed/energy. Neutron radiation consists of free neutrons. These neutrons may be emitted during either spontaneous or induced nuclear fission. Neutrons are rare radiation particles; they are produced in large numbers only where chain reaction fission or fusion reactions are active; this happens for about 10 microseconds in a thermonuclear explosion, or continuously inside an operating nuclear reactor; production of the neutrons stops almost immediately in the reactor when it goes non-critical.

Neutrons are the only type of ionizing radiation that can make other objects, or material, radioactive. This process, called neutron activation, is the primary method used to produce radioactive sources for use in medical, academic, and industrial applications. Even comparatively low speed thermal neutrons cause neutron activation (in fact, they cause it more efficiently). Neutrons do not ionize atoms in the same way that charged particles such as protons and electrons do (by the excitation of an electron), because neutrons have no charge. It is through their absorption by nuclei which then become unstable that they cause ionization. Hence, neutrons are said to be "indirectly ionizing." Even neutrons without significant kinetic energy are indirectly ionizing, and are thus a significant radiation hazard. Not all materials are capable of neutron activation; in water, for example, the most common isotopes of both types atoms present (hydrogen and oxygen) capture neutrons and become heavier but remain stable forms of those atoms. Only the absorption of more than one neutron, a statistically rare occurrence, can activate a hydrogen atom, while oxygen requires two additional absorptions. Thus water is only very weakly capable of activation. The sodium in salt (as in sea water), on the other hand, need only absorb a single neutron to become Na-24, a very intense source of beta decay, with half-life of 15 hours.

In addition, high-energy (high-speed) neutrons have the ability to directly ionize atoms. One mechanism by which high energy neutrons ionize atoms is to strike the nucleus of an atom and knock the atom out of a molecule, leaving one or more electrons behind as the chemical bond is broken. This leads to production of chemical free radicals. In addition, very high energy neutrons can cause ionizing radiation by "neutron spallation" or knockout, wherein neutrons cause emission of high-energy protons from atomic nuclei (especially hydrogen nuclei) on impact. The last process imparts most of the neutron's energy to the proton, much like one billiard ball striking another. The charged protons, and other products from such reactions are directly ionizing.

High-energy neutrons are very penetrating and can travel great distances in air (hundreds or even thousands of meters) and moderate distances (several meters) in common solids. They typically require hydrogen rich shielding, such as concrete or water, to block them within distances of less than a meter. A common source of neutron radiation occurs inside a nuclear reactor, where a meters-thick water layer is used as effective shielding.

Cosmic Radiation

There are two sources of high energy particles entering the Earth's atmosphere from outer space: the sun and deep space. The sun continuously emits particles, primarily free protons, in the solar wind, and occasionally augments the flow hugely with coronal mass ejections (CME).

The particles from deep space (inter- and extra-galactic) are much less frequent, but of much higher energies. These particles are also mostly protons, with much of the remainder consisting of helions (alpha particles). A few completely ionized nuclei of heavier elements are present. The origin of these galactic cosmic rays is not yet well understood, but they seem to be remnants of supernovae and especially gamma-ray bursts (GRB), which feature magnetic fields capable of the huge accelerations measured from these particles. They may also be generated by quasars, which are galaxy-wide jet phenomena similar to GRBs but known for their much larger size, and which seem to be a violent part of the universe's early history.

Non-ionizing Radiation

The kinetic energy of particles of non-ionizing radiation is too small to produce charged ions when passing through matter. For non-ionizing electromagnetic radiation, the associated particles (photons) have only sufficient energy to change the rotational, vibrational or electronic valence configurations of molecules and atoms. The effect of non-ionizing forms of radiation on living tissue has only recently been studied. Nevertheless, different biological effects are observed for different types of non-ionizing radiation.

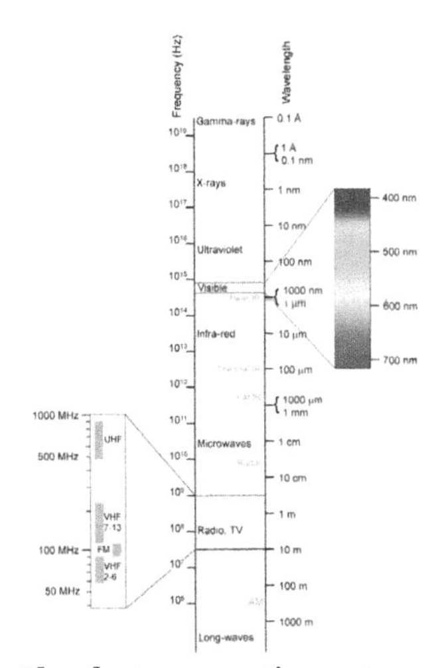

The electromagnetic spectrum

Even "non-ionizing" radiation is capable of causing thermal-ionization if it deposits enough heat to raise temperatures to ionization energies. These reactions occur at far higher energies than with ionization radiation, which requires only single particles to cause ionization. A familiar example of thermal ionization is the flame-ionization of a common fire, and the browning reactions in common food items induced by infrared radiation, during broiling-type cooking.

The electromagnetic spectrum is the range of all possible electromagnetic radiation frequencies. The electromagnetic spectrum (usually just spectrum) of an object is the characteristic distribution of electromagnetic radiation emitted by, or absorbed by, that particular object.

The non-ionizing portion of electromagnetic radiation consists of electromagnetic waves that (as individual quanta or particles) are not energetic enough to detach electrons from atoms or molecules and hence cause their ionization. These include radio waves, microwaves, infrared, and (sometimes) visible light. The lower frequencies of ultraviolet light may cause chemical changes and molecular damage similar to ionization, but is technically not ionizing. The highest frequencies of ultraviolet light, as well as all X-rays and gamma-rays are ionizing.

The occurrence of ionization depends on the energy of the individual particles or waves, and not on their number. An intense flood of particles or waves will not cause ionization if these particles or waves do not carry enough energy to be ionizing, unless they raise the temperature of a body to a point high enough to ionize small fractions of atoms or molecules by the process of thermal-ionization (this, however, requires relatively extreme radiation intensities).

Ultraviolet Light

As noted above, the lower part of the spectrum of ultraviolet, called soft UV, from 3 eV to about 10 eV, is non-ionizing. However, the effects of non-ionizing ultraviolet on chemistry and the damage to biological systems exposed to it (including oxidation, mutation, and cancer) are such that even this part of ultraviolet is often compared with ionizing radiation.

Visible Light

Light, or visible light, is a very narrow range of electromagnetic radiation of a wavelength that is visible to the human eye, or 380–750 nm which equates to a frequency range of 790 to 400 THz respectively. More broadly, physicists use the term "light" to mean electromagnetic radiation of all wavelengths, whether visible or not.

Infrared

Infrared (IR) light is electromagnetic radiation with a wavelength between 0.7 and 300 micrometers, which corresponds to a frequency range between 430 and 1 THz respectively. IR wavelengths are longer than that of visible light, but shorter than that of microwaves. Infrared may be detected at a distance from the radiating objects by "feel." Infrared sensing snakes can detect and focus infrared by use of a pinhole lens in their heads, called "pits". Bright sunlight provides an irradiance of just over 1 kilowatt per square meter at sea level. Of this energy, 53% is infrared radiation, 44% is visible light, and 3% is ultraviolet radiation.

Microwave

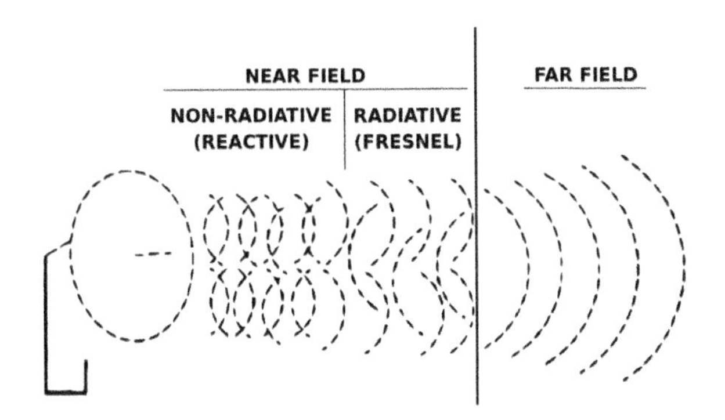

In electromagnetic radiation (such as microwaves from an antenna, shown here) the term "radiation" applies only to the parts of the electromagnetic field that radiate into infinite space and decrease in intensity by an inverse-square law of power so that the total radiation energy that crosses through an imaginary spherical surface is the same, no matter how far away from the antenna the spherical surface is drawn. Electromagnetic radiation includes the far field part of the electromagnetic field around a transmitter. A part of the "near-field" close to the transmitter, is part of the changing electromagnetic field, but does not count as electromagnetic radiation.

Microwaves are electromagnetic waves with wavelengths ranging from as short as one millimeter to as long as one meter, which equates to a frequency range of 300 MHz to 300 GHz. This broad definition includes both UHF and EHF (millimeter waves), but various sources use different other limits. In all cases, microwaves include the entire super high frequency band (3 to 30 GHz, or 10 to 1 cm) at minimum, with RF engineering often putting the lower boundary at 1 GHz (30 cm), and the upper around 100 GHz (3mm).

Radio Waves

Radio waves are a type of electromagnetic radiation with wavelengths in the electromagnetic spectrum longer than infrared light. Like all other electromagnetic waves, they travel at the speed of light. Naturally occurring radio waves are made by lightning, or by certain astronomical objects. Artificially generated radio waves are used for fixed and mobile radio communication, broadcasting, radar and other navigation systems, satellite communication, computer networks and innumerable other applications. In addition, almost any wire carrying alternating current will radiate some of the energy away as radio waves; these are mostly termed interference. Different frequencies of radio waves have different propagation characteristics in the Earth's atmosphere; long waves may bend at the rate of the curvature of the Earth and may cover a part of the Earth very consistently, shorter waves travel around the world by multiple reflections off the ionosphere and the Earth. Much shorter wavelengths bend or reflect very little and travel along the line of sight.

Very Low Frequency

Very low frequency (VLF) refers to a frequency range of 30 Hz to 3 kHz which corresponds to wavelengths of 100,000 to 10,000 meters respectively. Since there is not much bandwidth in this range of the radio spectrum, only the very simplest signals can be transmitted, such as for radio navigation. Also known as the myriameter band or myriameter wave as the wavelengths range from ten to one myriameter (an obsolete metric unit equal to 10 kilometers).

Extremely Low Frequency

Extremely low frequency (ELF) is radiation frequencies from 3 to 30 Hz (10^8 to 10^7 meters respectively). In atmosphere science, an alternative definition is usually given, from 3 Hz to 3 kHz. In the related magnetosphere science, the lower frequency electromagnetic oscillations (pulsations occurring below ~3 Hz) are considered to lie in the ULF range, which is thus also defined differently from the ITU Radio Bands. A massive military ELF antenna in Michigan radiates very slow messages to otherwise unreachable receivers, such as submerged submarines.

Thermal Radiation (Heat)

Thermal radiation is a common synonym for infrared radiation emitted by objects at temperatures often encountered on Earth. Thermal radiation refers not only to the radiation itself, but also the process by which the surface of an object radiates its thermal energy in the form black body radiation. Infrared or red radiation from a common household radiator or electric heater is an example of thermal radiation, as is the heat emitted by an operating incandescent light bulb. Thermal radiation is generated when energy from the movement of charged particles within atoms is converted to electromagnetic radiation.

As noted above, even low-frequency thermal radiation may cause temperature-ionization whenever it deposits sufficient thermal energy to raises temperatures to a high enough level. Common examples of this are the ionization (plasma) seen in common flames, and the molecular changes caused by the "browning" during food-cooking, which is a chemical process that begins with a large component of ionization.

Black-body Radiation

Black-body radiation is an idealized spectrum of radiation emitted by a body that is at a uniform temperature. The shape of the spectrum and the total amount of energy emitted by the body is a function the absolute temperature of that body. The radiation emitted covers the entire electromagnetic spectrum and the intensity of the radiation (power/unit-area) at a given frequency is described by Planck's law of radiation. For a given temperature of a black-body there is a particular frequency at which the radiation emitted is at its maximum intensity. That maximum radiation frequency moves toward higher frequencies as the temperature of the body increases. The frequency at which the black-body radiation is at maximum is given by Wien's displacement law and is a function of the body's absolute temperature. A black-body is one that emits at any temperature the maximum possible amount of radiation at any given wavelength. A black-body will also absorb the maximum possible incident radiation at any given wavelength. A black-body with a temperature at or below room temperature would thus appear absolutely black, as it would not reflect any incident light nor would it emit enough radiation at visible wavelengths for our eyes to detect. Theoretically, a black-body emits electromagnetic radiation over the entire spectrum from very low frequency radio waves to x-rays, creating a continuum of radiation.

The color of a radiating black-body tells the temperature of its radiating surface. It is responsible for the color of stars, which vary from infrared through red (2,500K), to yellow (5,800K), to white and to blue-white (15,000K) as the peak radiance passes through those points in the visible spectrum. When the peak is below the visible spectrum the body is black, while when it is above the body is blue-white, since all the visible colors are represented from blue decreasing to red.

Discovery

Electromagnetic radiation of wavelengths other than visible light were discovered in the early 19th century. The discovery of infrared radiation is ascribed to William Herschel, the astronomer. Herschel published his results in 1800 before the Royal Society of London. Herschel, like Ritter, used a prism to refract light from the Sun and detected the infrared (beyond the red part of the spectrum), through an increase in the temperature recorded by a thermometer.

In 1801, the German physicist Johann Wilhelm Ritter made the discovery of ultraviolet by noting that the rays from a prism darkened silver chloride preparations more quickly than violet light. Ritter's experiments were an early precursor to what would become photography. Ritter noted that the UV rays were capable of causing chemical reactions.

The first radio waves detected were not from a natural source, but were produced deliberately and artificially by the German scientist Heinrich Hertz in 1887, using electrical circuits calculated to produce oscillations in the radio frequency range, following formulas suggested by the equations of James Clerk Maxwell.

Wilhelm Röntgen discovered and named X-rays. While experimenting with high voltages applied to an evacuated tube on 8 November 1895, he noticed a fluorescence on a nearby plate of coated glass. Within a month, he discovered the main properties of X-rays that we understand to this day.

In 1896, Henri Becquerel found that rays emanating from certain minerals penetrated black paper and caused fogging of an unexposed photographic plate. His doctoral student Marie Curie discovered that only certain chemical elements gave off these rays of energy. She named this behavior radioactivity.

Alpha rays (alpha particles) and beta rays (beta particles) were differentiated by Ernest Rutherford through simple experimentation in 1899. Rutherford used a generic pitchblende radioactive source and determined that the rays produced by the source had differing penetrations in materials. One type had short penetration (it was stopped by paper) and a positive charge, which Rutherford named *alpha rays*. The other was more penetrating (able to expose film through paper but not metal) and had a negative charge, and this type Rutherford named *beta*. This was the radiation that had been first detected by Becquerel from uranium salts. In 1900, the French scientist Paul Villard discovered a third neutrally charged and especially penetrating type of radiation from radium, and after he described it, Rutherford realized it must be yet a third type of radiation, which in 1903 Rutherford named gamma rays.

Henri Becquerel himself proved that beta rays are fast electrons, while Rutherford and Thomas Royds proved in 1909 that alpha particles are ionized helium. Rutherford and Edward Andrade proved in 1914 that gamma rays are like X-rays, but with shorter wavelengths.

Cosmic ray radiations striking the Earth from outer space were finally definitively recognized and proven to exist in 1912, as the scientist Victor Hess carried an electrometer to various altitudes in a free balloon flight. The nature of these radiations was only gradually understood in later years.

Neutron radiation was discovered with the neutron by Chadwick, in 1932. A number of other high energy particulate radiations such as positrons, muons, and pions were discovered by cloud chamber examination of cosmic ray reactions shortly thereafter, and others types of particle radiation were produced artificially in particle accelerators, through the last half of the twentieth century.

Uses

Medicine

Radiation and radioactive substances are used for diagnosis, treatment, and research. X-rays, for example, pass through muscles and other soft tissue but are stopped by dense materials. This property of X-rays enables doctors to find broken bones and to locate cancers that might be growing in the body. Doctors also find certain diseases by injecting a radioactive substance and monitoring the radiation given off as the substance moves through the body. Radiation used for cancer treatment is called ionizing radiation because it forms ions in the cells of the tissues it passes through as it dislodges electrons from atoms. This can kill cells or change genes so the cells cannot grow. Other forms of radiation such as radio waves, microwaves, and light waves are called non-ionizing. They don't have as much energy and are not able to ionize cells.

Communication

All modern communication systems use forms of electromagnetic radiation. Variations in the intensity of the radiation represent changes in the sound, pictures, or other information being transmitted. For example, a human voice can be sent as a radio wave or microwave by making the wave vary to correspond variations in the voice. Musicians have also experimented with gamma sonification, or using nuclear radiation, to produce sound and music.

Science

Researchers use radioactive atoms to determine the age of materials that were once part of a living organism. The age of such materials can be estimated by measuring the amount of radioactive carbon they contain in a process called radiocarbon dating. Similarly, using other radioactive elements, the age of rocks and other geological features (even some man-made objects) can be determined; this is called Radiometric dating. Environmental scientists use radioactive atoms, known as tracer atoms, to identify the pathways taken by pollutants through the environment.

Radiation is used to determine the composition of materials in a process called neutron activation analysis. In this process, scientists bombard a sample of a substance with particles called neutrons. Some of the atoms in the sample absorb neutrons and become radioactive. The scientists can identify the elements in the sample by studying the emitted radiation.

References

- Sam Zhang; Dongliang Zhao (19 November 2012). Aeronautical and Aerospace Materials Handbook. CRC Press. pp. 304–. ISBN 978-1-4398-7329-8. Retrieved 7 May 2013.
- Martin Eein (2002). Drop-Surface Interactions. Springer. pp. 174–. ISBN 978-3-211-83692-7. Retrieved 7 May 2013.
- George E. Totten (2002). Handbook of Residual Stress and Deformation of Steel. ASM International. pp. 322–. ISBN 978-1-61503-227-3. Retrieved 7 May 2013.
- Bailyn, M. (1994). A Survey of Thermodynamics, American Institute of Physics, New York, ISBN 0-88318-797-3.
- Incropera DeWitt VBergham Lavine 2007, Introduction to Heat Transfer, 5th ed., pg. 6 ISBN 978-0-471-45727-5.

- Aroon Shenoy, Mikhail Sheremet, Ioan Pop, 2016, Convective Flow and Heat Transfer from Wavy Surfaces: Viscous Fluids, Porous Media, and Nanofluids, CRC Press, Taylor & Francis Group, Florida ISBN 978-1-498-76090-4.
- Rajiv Asthana; Ashok Kumar; Narendra B. Dahotre (9 January 2006). Materials Processing and Manufacturing Science. Butterworth–Heinemann. pp. 158–. ISBN 978-0-08-046488-6. Retrieved 7 May 2013.
- "Heat Transfer Mechanisms". Colorado State University. The College of Engineering at Colorado State University. Retrieved 14 September 2015.
- "ICRP Publication 103 The 2007 Recommendations of the International Commission on Protection" (PDF). ICRP. Retrieved 12 December 2013.

4

Essential Concepts of Heat Transfer

The property of a material to conduct heat is thermal conductivity whereas the change of shape and area of matter in response to temperature change through the phenomenon of heat conductivity is thermal expansion. Thermal equilibrium and its relation to heat transfer is also elucidated in the following chapter.

Thermal Expansion

Expansion joint in a road bridge used to avoid damage from thermal expansion.

Thermal expansion is the tendency of matter to change in shape, area, and volume in response to a change in temperature, through heat transfer.

Temperature is a monotonic function of the average molecular kinetic energy of a substance. When a substance is heated, the kinetic energy of its molecules increases. Thus, the molecules begin moving more and usually maintain a greater average separation. Materials which contract with increasing temperature are unusual; this effect is limited in size, and only occurs within limited temperature ranges. The degree of expansion divided by the change in temperature is called the material's coefficient of thermal expansion and generally varies with temperature.

Overview

Predicting Expansion

If an equation of state is available, it can be used to predict the values of the thermal expansion at all the required temperatures and pressures, along with many other state functions.

Contraction Effects (Negative Thermal Expansion)

A number of materials contract on heating within certain temperature ranges; this is usually called negative thermal expansion, rather than "thermal contraction". For example, the coefficient of thermal expansion of water drops to zero as it is cooled to 3.983 °C and then becomes negative below this temperature; this means that water has a maximum density at this temperature, and this leads to bodies of water maintaining this temperature at their lower depths during extended periods of sub-zero weather. Also, fairly pure silicon has a negative coefficient of thermal expansion for temperatures between about 18 and 120 Kelvin.

Factors Affecting Thermal Expansion

Unlike gases or liquids, solid materials tend to keep their shape when undergoing thermal expansion.

Thermal expansion generally decreases with increasing bond energy, which also has an effect on the melting point of solids, so, high melting point materials are more likely to have lower thermal expansion. In general, liquids expand slightly more than solids. The thermal expansion of glasses is higher compared to that of crystals. At the glass transition temperature, rearrangements that occur in an amorphous material lead to characteristic discontinuities of coefficient of thermal expansion or specific heat. These discontinuities allow detection of the glass transition temperature where a supercooled liquid transforms to a glass.

Absorption or desorption of water (or other solvents) can change the size of many common materials; many organic materials change size much more due to this effect than they do to thermal expansion. Common plastics exposed to water can, in the long term, expand by many percent.

Coefficient of Thermal Expansion

The coefficient of thermal expansion describes how the size of an object changes with a change in temperature. Specifically, it measures the fractional change in size per degree change in temperature at a constant pressure. Several types of coefficients have been developed: volumetric, area, and linear. Which is used depends on the particular application and which dimensions are considered important. For solids, one might only be concerned with the change along a length, or over some area.

The volumetric thermal expansion coefficient is the most basic thermal expansion coefficient, and the most relevant for fluids. In general, substances expand or contract when their temperature changes, with expansion or contraction occurring in all directions. Substances that expand at the same rate in every direction are called isotropic. For isotropic materials, the area and volumetric thermal expansion coefficient are, respectively, approximately twice and three times larger than the linear thermal expansion coefficient.

Mathematical definitions of these coefficients are defined below for solids, liquids, and gases.

General Volumetric Thermal Expansion Coefficient

In the general case of a gas, liquid, or solid, the volumetric coefficient of thermal expansion is given by

$$\alpha_V = \frac{1}{V}\left(\frac{\partial V}{\partial T}\right)_p$$

The subscript p indicates that the pressure is held constant during the expansion, and the subscript V stresses that it is the volumetric (not linear) expansion that enters this general definition. In the case of a gas, the fact that the pressure is held constant is important, because the volume of a gas will vary appreciably with pressure as well as temperature. For a gas of low density this can be seen from the ideal gas law.

Expansion in Solids

When calculating thermal expansion it is necessary to consider whether the body is free to expand or is constrained. If the body is free to expand, the expansion or strain resulting from an increase in temperature can be simply calculated by using the applicable coefficient of thermal expansion

If the body is constrained so that it cannot expand, then internal stress will be caused (or changed) by a change in temperature. This stress can be calculated by considering the strain that would occur if the body were free to expand and the stress required to reduce that strain to zero, through the stress/strain relationship characterised by the elastic or Young's modulus. In the special case of solid materials, external ambient pressure does not usually appreciably affect the size of an object and so it is not usually necessary to consider the effect of pressure changes.

Common engineering solids usually have coefficients of thermal expansion that do not vary significantly over the range of temperatures where they are designed to be used, so where extremely high accuracy is not required, practical calculations can be based on a constant, average, value of the coefficient of expansion.

Linear Expansion

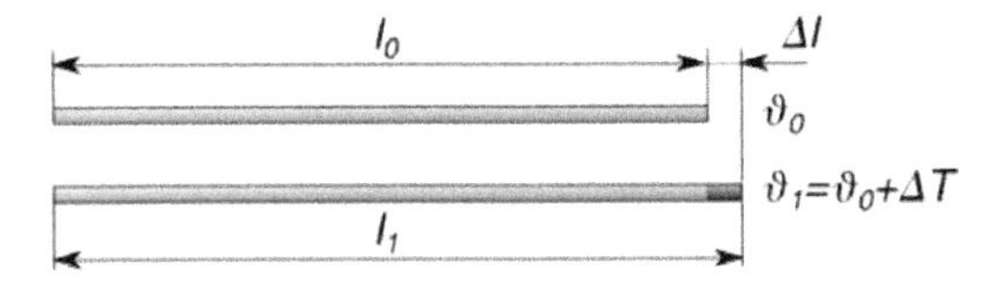

Change in length of a rod due to thermal expansion.

Linear expansion means change in one dimension (length) as opposed to change in volume (volumetric expansion). To a first approximation, the change in length measurements of an object due to thermal expansion is related to temperature change by a "linear expansion coefficient". It is the fractional change in length per degree of temperature change. Assuming negligible effect of pressure, we may write:

$$\alpha_L = \frac{1}{L}\frac{dL}{dT}$$

where L is a particular length measurement and dL / dT is the rate of change of that linear dimension per unit change in temperature.

The change in the linear dimension can be estimated to be:

$$\frac{\Delta L}{L} = \alpha_L \Delta T$$

This equation works well as long as the linear-expansion coefficient does not change much over the change in temperature ΔT, and the fractional change in length is small $\Delta L / L \ll 1$.. If either of these conditions does not hold, the equation must be integrated.

Effects on Strain

For solid materials with a significant length, like rods or cables, an estimate of the amount of thermal expansion can be described by the material strain, given by $\epsilon_{\text{thermal}}$ and defined as:

$$\epsilon_{\text{thermal}} = \frac{(L_{\text{final}} - L_{\text{initial}})}{L_{\text{initial}}}$$

where L_{initial} is the length before the change of temperature and L_{final} is the length after the change of temperature.

For most solids, thermal expansion is proportional to the change in temperature:

$$\epsilon_{\text{thermal}} \propto \Delta T$$

Thus, the change in either the strain or temperature can be estimated by:

$$\epsilon_{\text{thermal}} = \alpha_L \Delta T$$

where

$$\Delta T = (T_{\text{final}} - T_{\text{initial}})$$

is the difference of the temperature between the two recorded strains, measured in degrees Celsius or Kelvin, and is the linear coefficient of thermal expansion in "per degree Celsius" or "per Kelvin", denoted by $°C^{-1}$ or K^{-1}, respectively. In the field of continuum mechanics, the thermal expansion and its effects are treated as eigenstrain and eigenstress.

Area Expansion

The area thermal expansion coefficient relates the change in a material's area dimensions to a change in temperature. It is the fractional change in area per degree of temperature change. Ignoring pressure, we may write:

$$\alpha_A = \frac{1}{A}\frac{dA}{dT}$$

where A is some area of interest on the object, and dA / dT is the rate of change of that area per unit change in temperature.

The change in the area can be estimated as:

$$\frac{\Delta A}{A} = \alpha_A \Delta T$$

This equation works well as long as the area expansion coefficient does not change much over the change in temperature ΔT, and the fractional change in area is small $\Delta A / A \ll 1$. If either of these conditions does not hold, the equation must be integrated.

Volume Expansion

For a solid, we can ignore the effects of pressure on the material, and the volumetric thermal expansion coefficient can be written:

$$\alpha_V = \frac{1}{V}\frac{dV}{dT}$$

where V is the volume of the material, and dV / dT is the rate of change of that volume with temperature.

This means that the volume of a material changes by some fixed fractional amount. For example, a steel block with a volume of 1 cubic meter might expand to 1.002 cubic meters when the temperature is raised by 50 K. This is an expansion of 0.2%. If we had a block of steel with a volume of 2 cubic meters, then under the same conditions, it would expand to 2.004 cubic meters, again an expansion of 0.2%. The volumetric expansion coefficient would be 0.2% for 50 K, or 0.004% K^{-1}.

If we already know the expansion coefficient, then we can calculate the change in volume

$$\frac{\Delta V}{V} = \alpha_V \Delta T$$

where $\Delta V / V$ is the fractional change in volume (e.g., 0.002) and ΔT is the change in temperature (50 °C).

The above example assumes that the expansion coefficient did not change as the temperature changed and the increase in volume is small compared to the original volume. This is not always true, but for small changes in temperature, it is a good approximation. If the volumetric expansion coefficient does change appreciably with temperature, or the increase in volume is significant, then the above equation will have to be integrated:

$$\ln\left(\frac{V+\Delta V}{V}\right)=\int_{T_i}^{T_f}\alpha_V(T)dT$$

$$\frac{\Delta V}{V}=\exp\left(\int_{T_i}^{T_f}\alpha_V(T)dT\right)-1$$

where $\alpha_V(T)$ is the volumetric expansion coefficient as a function of temperature T, and T_i, T_f are the initial and final temperatures respectively.

Isotropic Materials

For isotropic materials the volumetric thermal expansion coefficient is three times the linear coefficient:

$$\alpha_V=3\alpha_L$$

This ratio arises because volume is composed of three mutually orthogonal directions. Thus, in an isotropic material, for small differential changes, one-third of the volumetric expansion is in a single axis. As an example, take a cube of steel that has sides of length L. The original volume will be $V=L^3$ and the new volume, after a temperature increase, will be

$$V+\Delta V=(L+\Delta L)^3=L^3+3L^2\Delta L+3L\Delta L^2+\Delta L^3\approx L^3+3L^2\Delta L=V+3V\frac{\Delta L}{L}$$

We can make the substitutions $\Delta V=\alpha_V L^3\Delta T$ and, for isotropic materials, $\Delta L=\alpha_L L\Delta T$. We now have:

$$V+\Delta V=(L+L\alpha_L\Delta T)^3=L^3+3L^3\alpha_L\Delta T+3L^3\alpha_L^2\Delta T^2+L^3\alpha_L^3\Delta T^3\approx L^3+3L^3\alpha_L\Delta T$$

Since the volumetric and linear coefficients are defined only for extremely small temperature and dimensional changes (that is, when ΔT and ΔL are small), the last two terms can be ignored and we get the above relationship between the two coefficients. If we are trying to go back and forth between volumetric and linear coefficients using larger values of ΔT then we will need to take into account the third term, and sometimes even the fourth term.

Similarly, the area thermal expansion coefficient is two times the linear coefficient:

$$\alpha_A=2\alpha_L$$

This ratio can be found in a way similar to that in the linear example above, noting that the area of a face on the cube is just L^2 . Also, the same considerations must be made when dealing with large values of ΔT .

Anisotropic Materials

Materials with anisotropic structures, such as crystals (with less than cubic symmetry) and many composites, will generally have different linear expansion coefficients $-\alpha_L$ in different directions. As a result, the total volumetric expansion is distributed unequally among the three axes. If the crystal symmetry is monoclinic or triclinic, even the angles between these axes are subject to thermal changes. In such cases it is necessary to treat the coefficient of thermal expansion as a tensor with up to six independent elements. A good way to determine the elements of the tensor is to study the expansion by powder diffraction.

Isobaric Expansion in Gases

For an ideal gas, the volumetric thermal expansion (i.e., relative change in volume due to temperature change) depends on the type of process in which temperature is changed. Two simple cases are isobaric change, where pressure is held constant, and adiabatic change, where no heat is exchanged with the environment.

The ideal gas law can be written as:

$$pv = T$$

where p is the pressure, v is the specific volume, and t is temperature measured in *energy units*. By taking the logarithm of this equation:

$$\ln(v) + \ln(p) = \ln(T)$$

Then by definition of isobaric thermal expansion coefficient, with the above equation of state:

$$\gamma_p \equiv \frac{1}{v}\left(\frac{\partial v}{\partial T}\right)_p = \left(\frac{d(\ln v)}{dT}\right)_p = \frac{d(\ln T)}{dT} = \frac{1}{T}.$$

The index p denotes an isobaric process.

Expansion in Liquids

Theoretically, the coefficient of linear expansion can be found from the coefficient of volumetric expansion ($\alpha_V \approx 3\alpha$). However, for liquids, α is calculated through the experimental determination of α_V.

Expansion in mixtures and alloys

The expansivity of the components of the mixture can cancel each other like in invar.

The thermal expansivity of a mixture from the expansivities of the pure components and their excess expansivities follow from:

$$\frac{\partial V}{\partial T} = \sum_i \frac{\partial V_i}{\partial T} + \sum_i \frac{\partial V_i^E}{\partial T}$$

$$\alpha = \sum_i \alpha_i V_i + \sum_i \alpha_i^E V_i^E$$

$$\frac{\partial \bar{V}^E{}_i}{\partial T} = R\frac{\partial(ln(\gamma_i))}{\partial P} + RT\frac{\partial^2}{\partial T \partial P} ln(\gamma_i)$$

Apparent and Absolute Expansion

When measuring the expansion of a liquid, the measurement must account for the expansion of the container as well. For example, a flask that has been constructed with a long narrow stem filled with enough liquid that the stem itself is partially filled, when placed in a heat bath will initially show the column of liquid in the stem to drop followed by the immediate increase of that column until the flask-liquid-heat bath system has thermalized. The initial observation of the column of liquid dropping is not due to an initial contraction of the liquid but rather the expansion of the flask as it contacts the heat bath first. Soon after, the liquid in the flask is heated by the flask itself and begins to expand. Since liquids typically have a greater expansion over solids, the liquid in the flask eventually exceeds that of the flask, causing the column of liquid in the flask to rise. A direct measurement of the height of the liquid column is a measurement of the apparent expansion of the liquid. The absolute expansion of the liquid is the apparent expansion corrected for the expansion of the containing vessel.

Examples and Applications

Thermal expansion of long continuous sections of rail tracks is the driving force for rail buckling. This phenomenon resulted in 190 train derailments during 1998–2002 in the US alone.

The expansion and contraction of materials must be considered when designing large structures, when using tape or chain to measure distances for land surveys, when designing molds for casting hot material, and in other engineering applications when large changes in dimension due to temperature are expected.

Thermal expansion is also used in mechanical applications to fit parts over one another, e.g. a bushing can be fitted over a shaft by making its inner diameter slightly smaller than the diameter of the shaft, then heating it until it fits over the shaft, and allowing it to cool after it has been pushed over the shaft, thus achieving a 'shrink fit'. Induction shrink fitting is a common industrial method to pre-heat metal components between 150 °C and 300 °C thereby causing them to expand and allow for the insertion or removal of another component.

There exist some alloys with a very small linear expansion coefficient, used in applications that demand very small changes in physical dimension over a range of temperatures. One of these is Invar 36, with α approximately equal to 0.6×10^{-6} K^{-1}. These alloys are useful in aerospace applications where wide temperature swings may occur.

Pullinger's apparatus is used to determine the linear expansion of a metallic rod in the laboratory. The apparatus consists of a metal cylinder closed at both ends (called a steam jacket). It is provided with an inlet and outlet for the steam. The steam for heating the rod is supplied by a boiler which is connected by a rubber tube to the inlet. The center of the cylinder contains a hole to insert a thermometer. The rod under investigation is enclosed in a steam jacket. One of its ends is free, but the other end is pressed against a fixed screw. The position of the rod is determined by a micrometer screw gauge or spherometer.

Drinking glass with fracture due to uneven thermal expansion after pouring of hot liquid into the otherwise cool glass

The control of thermal expansion in brittle materials is a key concern for a wide range of reasons. For example, both glass and ceramics are brittle and uneven temperature causes uneven expansion which again causes thermal stress and this might lead to fracture. Ceramics need to be joined or work in consort with a wide range of materials and therefore their expansion must be matched to the application. Because glazes need to be firmly attached to the underlying porcelain (or other body type) their thermal expansion must be tuned to 'fit' the body so that crazing or shivering do not occur. Good example of products whose thermal expansion is the key to their success are CorningWare and the spark plug. The thermal expansion of ceramic bodies can be controlled by firing to create crystalline species that will influence the overall expansion of the material in the desired direction. In addition or instead the formulation of the body can employ materials delivering particles of the desired expansion to the matrix. The thermal expansion of glazes is controlled by their chemical composition and the firing schedule to which they were subjected. In most cases there are complex issues involved in controlling body and glaze expansion, adjusting for thermal expansion must be done with an eye to other properties that will be affected, generally trade-offs are required.

Thermal expansion can have a noticeable effect in gasoline stored in above ground storage tanks which can cause gasoline pumps to dispense gasoline which may be more compressed than gasoline held in underground storage tanks in the winter time or less compressed than gasoline held in underground storage tanks in the summer time.

Heat-induced expansion has to be taken into account in most areas of engineering. A few examples are:

- Metal framed windows need rubber spacers
- Rubber tires
- Metal hot water heating pipes should not be used in long straight lengths
- Large structures such as railways and bridges need expansion joints in the structures to avoid sun kink
- One of the reasons for the poor performance of cold car engines is that parts have inefficiently large spacings until the normal operating temperature is achieved.
- A gridiron pendulum uses an arrangement of different metals to maintain a more temperature stable pendulum length.
- A power line on a hot day is droopy, but on a cold day it is tight. This is because the metals expand under heat.
- Expansion joints that absorb the thermal expansion in a piping system.
- Precision engineering nearly always requires the engineer to pay attention to the thermal expansion of the product. For example, when using a scanning electron microscope even small changes in temperature such as 1 degree can cause a sample to change its position relative to the focus point.

Thermometers are another application of thermal expansion – most contain a liquid (usually mercury or alcohol) which is constrained to flow in only one direction (along the tube) due to changes in volume brought about by changes in temperature. A bi-metal mechanical thermometer uses a bimetallic strip and bends due to the differing thermal expansion of the two metals.

Metal pipes made of different materials are heated by passing steam through them. While each pipe is being tested, one end is securely fixed and the other rests on a rotating shaft, the motion of which is indicated with a pointer. The linear expansion of the different metals is compared qualitatively and the coefficient of linear thermal expansion is calculated.

Thermal Expansion Coefficients for Various Materials

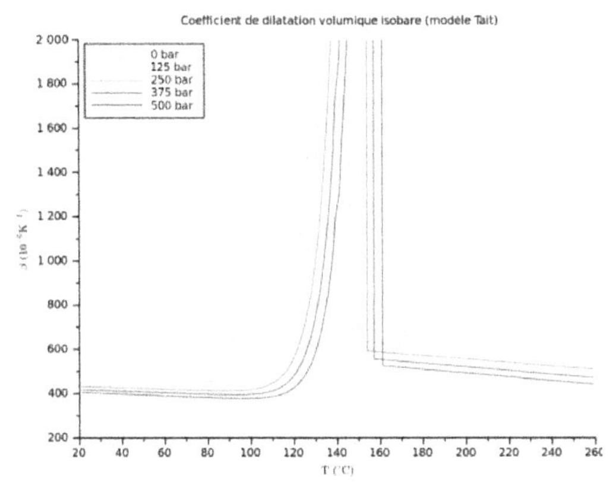

Volumetric thermal expansion coefficient for a semicrystalline polypropylene.

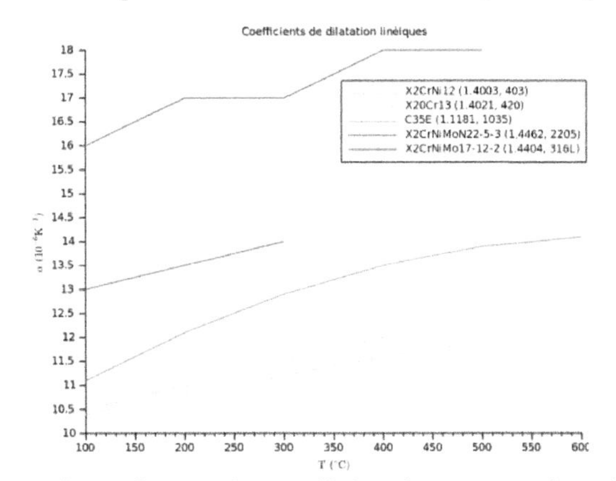

Linear thermal expansion coefficient for some steel grades.

This section summarizes the coefficients for some common materials.

For isotropic materials the coefficients linear thermal expansion α and volumetric thermal expansion α_V are related by $\alpha_V = 3\alpha$. For liquids usually the coefficient of volumetric expansion is listed and linear expansion is calculated here for comparison.

For common materials like many metals and compounds, the thermal expansion coefficient is inversely proportional to the melting point. In particular for metals the relation is:

$$\alpha \approx \frac{0.020}{M_P}$$

for halides and oxides

$$\alpha \approx \frac{0.038}{M_P} - 7.0 \cdot 10^{-6}\,\mathrm{K}^{-1}$$

In the table below, the range for α is from 10^{-7} K^{-1} for hard solids to 10^{-3} K^{-1} for organic liquids. The coefficient α varies with the temperature and some materials have a very high variation ; see for example the variation vs. temperature of the volumetric coefficient for a semicrystalline polypropylene (PP) at different pressure, and the variation of the linear coefficient vs. temperature for some steel grades (from bottom to top: ferritic stainless steel, martensitic stainless steel, carbon steel, duplex stainless steel, austenitic steel).

(The formula $\alpha_V \approx 3\alpha$ is usually used for solids.)

Material	Linear coefficient α at 20 °C (10^{-6} K^{-1})	Volumetric coefficient α_V at 20 °C (10^{-6} K^{-1})	Notes
Aluminium	23.1	69	
Aluminium nitride	5.3	4.2	
Benzocyclobutene	42	126	
Brass	19	57	
Carbon steel	10.8	32.4	
CFRP	– 0.8	Anisotropic	Fiber direction
Concrete	12	36	
Copper	17	51	
Diamond	1	3	
Ethanol	250	750	
Gallium(III) arsenide	5.8	17.4	
Gasoline	317	950	
Glass	8.5	25.5	
Glass, borosilicate	3.3	9.9	matched sealing partner for tungsten, molybdenum and kovar.
Glass (Pyrex)	3.2		
Glycerine		485	
Gold	14	42	
Helium		36.65	
Indium phosphide	4.6	13.8	
Invar	1.2	3.6	
Iron	11.8	33.3	
Kapton	20	60	DuPont Kapton 200EN

Lead	29	87	
Macor	9.3		
Magnesium	26	78	
Mercury	61	182	
Molybdenum	4.8	14.4	
Nickel	13	39	
Oak	54		Perpendicular to the grain
Douglas-fir	27	75	radial
Douglas-fir	45	75	tangential
Douglas-fir	3.5	75	parallel to grain
Platinum	9	27	
PP	150	450	
PVC	52	156	
Quartz (fused)	0.59	1.77	
Quartz	0.33	1	
Rubber	*disputed*	*disputed*	
Sapphire	5.3		Parallel to C axis, or
Silicon Carbide	2.77	8.31	
Silicon	2.56	9	
Silver	18	54	
Sitall	0±0.15	0±0.45	average for −60 °C to 60 °C
Stainless steel	10.1 ~ 17.3	51.9	
Steel	11.0 ~ 13.0	33.0 ~ 39.0	Depends on composition
Titanium	8.6	26	
Tungsten	4.5	13.5	
Turpentine		90	
Water	69	207	
YbGaGe	=0	=0	Refuted
Zerodur	≈0.02		at 0...50 °C

Thermal Conductivity

In physics, thermal conductivity (often denoted k, λ, or κ) is the property of a material to conduct heat. It is evaluated primarily in terms of Fourier's Law for heat conduction.

Heat transfer occurs at a lower rate across materials of low thermal conductivity than across materials of high thermal conductivity. Correspondingly, materials of high thermal conductivity are widely used in heat sink applications and materials of low thermal conductivity are used as thermal insulation. The thermal conductivity of a material may depend on temperature. The reciprocal of thermal conductivity is called thermal resistivity.

Thermal conductivity is actually a tensor, which means it is possible to have different values in different directions.

Units of Thermal Conductivity

In SI units, thermal conductivity is measured in watts per meter kelvin (W/(m·K)). The dimension of thermal conductivity is $M^1L^1T^{-3}\Theta^{-1}$. These variables are mass (M) mass, length (L), time (T), and temperature (Θ). In Imperial units, thermal conductivity is measured in BTU/(hr·ft·°F).

Other units which are closely related to the thermal conductivity are in common use in the construction and textile industries. The construction industry makes use of units such as the R-value (resistance) and the U-value (conductivity). Although related to the thermal conductivity of a material used in an insulation product, R- and U-values are dependent on the thickness of the product.

Likewise the textile industry has several units including the tog and the clo which express thermal resistance of a material in a way analogous to the R-values used in the construction industry.

Measurement

There are a number of ways to measure thermal conductivity. Each of these is suitable for a limited range of materials, depending on the thermal properties and the medium temperature. There is a distinction between steady-state and transient techniques.

In general, steady-state techniques are useful when the temperature of the material does not change with time. This makes the signal analysis straightforward (steady state implies constant signals). The disadvantage is that a well-engineered experimental set-up is usually needed. The Divided Bar (various types) is the most common device used for consolidated rock solids.

Experimental Values

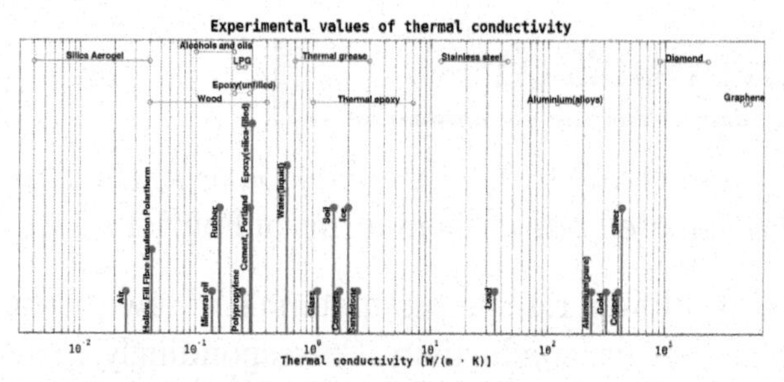

Experimental values of thermal conductivity.

Thermal conductivity is important in material science, research, electronics, building insulation and related fields, especially where high operating temperatures are achieved. Several materials are shown in the list of thermal conductivities. These should be considered approximate due to the uncertainties related to material definitions.

High energy generation rates within electronics or turbines require the use of materials with high thermal conductivity such as copper, aluminium, and silver. On the other hand, materials with low thermal conductance, such as poly-styrene and alumina, are used in building construction or in furnaces in an effort to slow the flow of heat, i.e. for insulation purposes.

Definitions

The reciprocal of thermal conductivity is *thermal resistivity*, usually expressed in kelvin-meters per watt ($K \cdot m \cdot W^{-1}$). For a given thickness of a material, that particular construction's *thermal resistance* and the reciprocal property, *thermal conductance*, can be calculated. Unfortunately, there are differing definitions for these terms.

Thermal conductivity, k, often depends on temperature. Therefore, the definitions listed below make sense when the thermal conductivity is temperature independent. Otherwise an representative mean value has to be considered.

Conductance

For general scientific use, *thermal conductance* is the quantity of heat that passes in unit time through a plate of *particular area and thickness* when its opposite faces differ in temperature by one kelvin. For a plate of thermal conductivity k, area A and thickness L, the conductance calculated is kA/L, measured in $W \cdot K^{-1}$ (equivalent to: W/°C). The thermal conductance of that particular construction is the inverse of the thermal resistance. Thermal conductivity and conductance are analogous to electrical conductivity ($A \cdot m^{-1} \cdot V^{-1}$) and electrical conductance ($A \cdot V^{-1}$).

There is also a measure known as heat transfer coefficient: the quantity of heat that passes in unit time through a *unit area* of a plate of particular thickness when its opposite faces differ in temperature by one kelvin. The reciprocal is *thermal insulance*. In summary:

- thermal conductance = kA/L, measured in $W \cdot K^{-1}$
    - thermal resistance = $L/(kA)$, measured in $K \cdot W^{-1}$ (equivalent to: °C/W)
- heat transfer coefficient = k/L, measured in $W \cdot K^{-1} \cdot m^{-2}$
    - thermal insulance = L/k, measured in $K \cdot m^2 \cdot W^{-1}$.

The heat transfer coefficient is also known as *thermal admittance* in the sense that the material may be seen as admitting heat to flow.

Resistance

Thermal resistance is the ability of a material to resist the flow of heat.

Thermal resistance is the reciprocal of thermal conductance, i.e., lowering its value will raise the heat conduction and vice versa.

When thermal resistances occur in series, they are *additive.* Thus, when heat flows consecutively through two components each with a resistance of 3 °C/W, the total resistance is 3+3=6 °C/W.

A common engineering design problem involves the selection of an appropriate sized heat sink for a given heat source. Working in units of thermal resistance greatly simplifies the design calculation. The following formula can be used to estimate the performance:

$$R_{hs} = \frac{\Delta T}{P_{th}} - R_s$$

where:

- R_{hs} is the maximum thermal resistance of the heat sink to ambient, in °C/W (equivalent to K/W)
- ΔT is the required temperature difference (temperature drop), in °C
- P_{th} is the thermal power (heat flow), in watts
- R_s is the thermal resistance of the heat source, in °C/W

For example, if a component produces 100 W of heat, and has a thermal resistance of 0.5 °C/W, what is the maximum thermal resistance of the heat sink? Suppose the maximum temperature is 125 °C, and the ambient temperature is 25 °C; then ΔT is 100 °C. The heat sink's thermal resistance to ambient must then be 0.5 °C/W or less (total resistance component and heat sink is then 1.0 °C/W).

Transmittance

A third term, *thermal transmittance,* quantifies the thermal conductance of a structure along with heat transfer due to convection and radiation. It is measured in the same units as thermal conductance and is sometimes known as the *composite thermal conductance.* The term *U-value* is often used.

Admittance

The thermal admittance of a material, such as a building fabric, is a measure of the ability of a material to transfer heat in the presence of a temperature difference on opposite sides of the material. Thermal admittance is measured in the same units as a heat transfer coefficient, power (watts) per unit area (square meters) per temperature change (kelvin). Thermal admittance of a building fabric affects a building's thermal response to variation in outside temperature.

Influencing Factors

Temperature

The effect of temperature on thermal conductivity is different for metals and non-metals. In metals conductivity is primarily due to free electrons. Following the Wiedemann–Franz law, thermal conductivity of metals is approximately proportional to the absolute temperature (in kelvin) times electrical conductivity. In pure metals the electrical conductivity decreases with increasing temperature and thus the product of the two, the thermal conductivity, stays approximately constant. In alloys the change in electrical conductivity is usually smaller and thus thermal conductivity increases with temperature, often proportionally to temperature.

On the other hand, heat conductivity in nonmetals is mainly due to lattice vibrations (phonons). Except for high quality crystals at low temperatures, the phonon mean free path is not reduced significantly at higher temperatures. Thus, the thermal conductivity of nonmetals is approximately constant at high temperatures. At low temperatures well below the Debye temperature, thermal conductivity decreases, as does the heat capacity.

Chemical Phase

When a material undergoes a phase change from solid to liquid or from liquid to gas the thermal conductivity may change. An example of this would be the change in thermal conductivity that occurs when ice (thermal conductivity of 2.18 W/(m·K) at 0 °C) melts to form liquid water (thermal conductivity of 0.56 W/(m·K) at 0 °C).

Thermal Anisotropy

Some substances, such as non-cubic crystals, can exhibit different thermal conductivities along different crystal axes, due to differences in phonon coupling along a given crystal axis. Sapphire is a notable example of variable thermal conductivity based on orientation and temperature, with 35 W/(m·K) along the C-axis and 32 W/(m·K) along the A-axis. Wood generally conducts better along the grain than across it.

When anisotropy is present, the direction of heat flow may not be exactly the same as the direction of the thermal gradient.

Electrical Conductivity

In metals, thermal conductivity approximately tracks electrical conductivity according to the Wiedemann–Franz law, as freely moving valence electrons transfer not only electric current but also heat energy. However, the general correlation between electrical and thermal conductance does not hold for other materials, due to the increased importance of phonon carriers for heat in non-metals. Highly electrically conductive silver is less thermally conductive than diamond, which is an electrical insulator, but due to its orderly array of atoms it is conductive of heat via phonons.

Magnetic Field

The influence of magnetic fields on thermal conductivity is known as the Righi-Leduc effect.

Convection

Air and other gases are generally good insulators, in the absence of convection. Therefore, many insulating materials function simply by having a large number of gas-filled pockets which prevent large-scale convection. Examples of these include expanded and extruded polystyrene (popularly referred to as "styrofoam") and silica aerogel, as well as warm clothes. Natural, biological insulators such as fur and feathers achieve similar effects by dramatically inhibiting convection of air or water near an animal's skin.

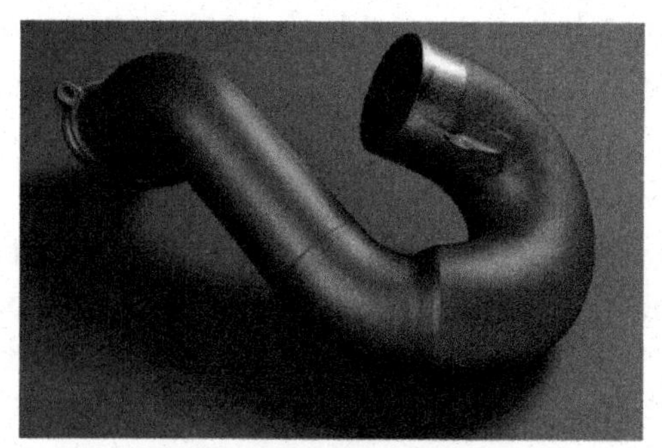

Ceramic coatings with low thermal conductivities are used on exhaust systems to prevent heat from reaching sensitive components

Light gases, such as hydrogen and helium typically have high thermal conductivity. Dense gases such as xenon and dichlorodifluoromethane have low thermal conductivity. An exception, sulfur hexafluoride, a dense gas, has a relatively high thermal conductivity due to its high heat capacity. Argon, a gas denser than air, is often used in insulated glazing (double paned windows) to improve their insulation characteristics.

Physical Origins

Heat flux is exceedingly difficult to control and isolate in a laboratory setting. At the atomic level, there are no simple, correct expressions for thermal conductivity. Atomically, the thermal conductivity of a system is determined by how atoms composing the system interact. There are two different approaches for calculating the thermal conductivity of a system.

- The first approach employs the Green-Kubo relations. Although this employs analytic expressions, which, in principle, can be solved, calculating the thermal conductivity of a dense fluid or solid using this relation requires the use of molecular dynamics computer simulation.
- The second approach is based on the relaxation time approach. Due to the anharmonicity within the crystal potential, the phonons in the system are known to scatter. There are three main mechanisms for scattering:
    - Boundary scattering, a phonon hitting the boundary of a system;
    - Mass defect scattering, a phonon hitting an impurity within the system and scattering;
    - Phonon-phonon scattering, a phonon breaking into two lower energy phonons or a phonon colliding with another phonon and merging into one higher-energy phonon.

Lattice Waves

Heat transport in both amorphous and crystalline dielectric solids is by way of elastic vibrations of the lattice (phonons). This transport mode is limited by the elastic scattering of acoustic phonons at lattice defects. These predictions were confirmed by the experiments of Chang and Jones on commercial glasses and glass ceramics, where the mean free paths were limited by "internal boundary scattering" to length scales of 10^{-2} cm to 10^{-3} cm.

The phonon mean free path has been associated directly with the effective relaxation length for processes without directional correlation. If V_g is the group velocity of a phonon wave packet, then the relaxation length l is defined as:

$$l = V_g t$$

where t is the characteristic relaxation time. Since longitudinal waves have a much greater phase velocity than transverse waves, V_{long} is much greater than V_{trans}, and the relaxation length or mean free path of longitudinal phonons will be much greater. Thus, thermal conductivity will be largely determined by the speed of longitudinal phonons.

Regarding the dependence of wave velocity on wavelength or frequency (dispersion), low-frequency phonons of long wavelength will be limited in relaxation length by elastic Rayleigh scattering. This type of light scattering from small particles is proportional to the fourth power of the frequency. For higher frequencies, the power of the frequency will decrease until at highest frequencies scattering is almost frequency independent.

Similar arguments were subsequently generalized to many glass forming substances using Brillouin scattering.

Phonons in the acoustical branch dominate the phonon heat conduction as they have greater energy dispersion and therefore a greater distribution of phonon velocities. Additional optical modes could also be caused by the presence of internal structure (i.e., charge or mass) at a lattice point; it is implied that the group velocity of these modes is low and therefore their contribution to the lattice thermal conductivity λ_L (κ_L) is small.

Each phonon mode can be split into one longitudinal and two transverse polarization branches. By extrapolating the phenomenology of lattice points to the unit cells it is seen that the total number of degrees of freedom is 3pq when p is the number of primitive cells with q atoms/unit cell. From these only 3p are associated with the acoustic modes, the remaining 3p(q-1) are accommodated through the optical branches. This implies that structures with larger p and q contain a greater number of optical modes and a reduced λ_L.

From these ideas, it can be concluded that increasing crystal complexity, which is described by a complexity factor CF (defined as the number of atoms/primitive unit cell), decreases λ_L. Micheline Roufosse and P.G. Klemens derived the exact proportionality in their article Thermal Conductivity of Complex Dielectric Crystals at Phys. Rev. B 7, 5379–5386 (1973). This was done by assuming that the relaxation time τ decreases with increasing number of atoms in the unit cell and then scaling the parameters of the expression for thermal conductivity in high temperatures accordingly.

Describing of anharmonic effects is complicated because exact treatment as in the harmonic case is not possible and phonons are no longer exact eigensolutions to the equations of motion. Even if the state of motion of the crystal could be described with a plane wave at a particular time, its accuracy would deteriorate progressively with time. Time development would have to be described by introducing a spectrum of other phonons, which is known as the phonon decay. The two most important anharmonic effects are the thermal expansion and the phonon thermal conductivity.

Only when the phonon number $\langle n \rangle$ deviates from the equilibrium value $\langle n \rangle^0$, can a thermal current arise as stated in the following expression

$$Q_x = \frac{1}{V}\sum_{q,j} \hbar\omega\left(\langle n \rangle - \langle n \rangle^0\right) v_x,$$

where v is the energy transport velocity of phonons. Only two mechanisms exist that can cause time variation of $\langle n \rangle$ in a particular region. The number of phonons that diffuse into the region from neighboring regions differs from those that diffuse out, or phonons decay inside the same region into other phonons. A special form of the Boltzmann equation

$$\frac{d\langle n\rangle}{dt}=\left(\frac{\partial\langle n\rangle}{\partial t}\right)_{\text{diff.}}+\left(\frac{\partial\langle n\rangle}{\partial t}\right)_{\text{decay}}$$

states this. When steady state conditions are assumed the total time derivate of phonon number is zero, because the temperature is constant in time and therefore the phonon number stays also constant. Time variation due to phonon decay is described with a relaxation time (τ) approximation

$$\left(\frac{\partial\langle n\rangle}{\partial t}\right)_{\text{decay}}=-\frac{\langle n\rangle-\langle n\rangle^{0}}{\tau},$$

which states that the more the phonon number deviates from its equilibrium value, the more its time variation increases. At steady state conditions and local thermal equilibrium are assumed we get the following equation

$$\left(\frac{\partial(n)}{\partial t}\right)_{\text{diff.}}=-v_x\frac{\partial(n)^{0}}{\partial T}\frac{\partial T}{\partial x}.$$

Using the relaxation time approximation for the Boltzmann equation and assuming steady-state conditions, the phonon thermal conductivity λ_L can be determined. The temperature dependence for λ_L originates from the variety of processes, whose significance for λ_L depends on the temperature range of interest. Mean free path is one factor that determines the temperature dependence for λ_L, as stated in the following equation

$$\lambda_L=\frac{1}{3V}\sum_{q,j}v(q,j)\Lambda(q,j)\frac{\partial}{\partial T}\epsilon(\omega(q,j),T),$$

where Λ is the mean free path for phonon and $\frac{\partial}{\partial T}\epsilon$ denotes the heat capacity. This equation is a result of combining the four previous equations with each other and knowing that $\langle v_x^2\rangle=\frac{1}{3}v^2$ for cubic or isotropic systems and $\Lambda=v\tau$.

At low temperatures (<10 K) the anharmonic interaction does not influence the mean free path and therefore, the thermal resistivity is determined only from processes for which q-conservation does not hold. These processes include the scattering of phonons by crystal defects, or the scattering from the surface of the crystal in case of high quality single crystal. Therefore, thermal conductance depends on the external dimensions of the crystal and the quality of the surface. Thus, temperature dependence of λ_L is determined by the specific heat and is therefore proportional to T^3.

Phonon quasimomentum is defined as ħq and differs from normal momentum because it is only defined within an arbitrary reciprocal lattice vector. At higher temperatures (10 K<T <Θ), the conservation of energy $\hbar\omega_1 = \hbar\omega_2 + \hbar\omega_3$ and quasimomentum $q_1 = q_2 + q_3 + G$, where q_1 is wave vector of the incident phonon and q_2, q_3 are wave vectors of the resultant phonons, may also involve a reciprocal lattice vector G complicating the energy transport process. These processes can also reverse the direction of energy transport.

Therefore, these processes are also known as Umklapp (U) processes and can only occur when phonons with sufficiently large q-vectors are excited, because unless the sum of q_2 and q_3 points outside of the Brillouin zone the momentum is conserved and the process is normal scattering (N-process). The probability of a phonon to have energy E is given by the Boltzmann distribution $P \propto e^{-E/kT}$. To U-process to occur the decaying phonon to have a wave vector q_1 that is roughly half of the diameter of the Brillouin zone, because otherwise quasimomentum would not be conserved.

Therefore, these phonons have to possess energy of $\sim k\Theta/2$, which is a significant fraction of Debye energy that is needed to generate new phonons. The probability for this is proportional to $e^{-\Theta/bT}$, with $b = 2$. Temperature dependence of the mean free path has an exponential form $e^{\Theta/bT}$. The presence of the reciprocal lattice wave vector implies a net phonon backscattering and a resistance to phonon and thermal transport resulting finite λ_L, as it means that momentum is not conserved. Only momentum non-conserving processes can cause thermal resistance.

At high temperatures (T>Θ) the mean free path and therefore λ_L has a temperature dependence T^{-1}, to which one arrives from formula $e^{\Theta/bT}$ by making the following ap proximation $e^x \propto x, (x) < 1$ and writing $x = \Theta/bT$. This dependency is known as Eucken's law and originates from the temperature dependency of the probability for the U-process to occur.

Thermal conductivity is usually described by the Boltzmann equation with the relaxation time approximation in which phonon scattering is a limiting factor. Another approach is to use analytic models or molecular dynamics or Monte Carlo based methods to describe thermal conductivity in solids.

Short wavelength phonons are strongly scattered by impurity atoms if an alloyed phase is present, but mid and long wavelength phonons are less affected. Mid and long wavelength phonons carry significant fraction of heat, so to further reduce lattice thermal conductivity one has to introduce structures to scatter these phonons. This is achieved by introducing interface scattering mechanism, which requires structures whose characteristic length is longer than that of impurity atom. Some possible ways to realize these interfaces are nanocomposites and embedded nanoparticles/structures.

Electronic Thermal Conductivity

Hot electrons from higher energy states carry more thermal energy than cold electrons, while electrical conductivity is rather insensitive to the energy distribution of carriers because the amount of charge that electrons carry, does not depend on their energy. This is a physical reason for the greater sensitivity of electronic thermal conductivity to energy dependence of density of states and relaxation time, respectively.

Mahan and Sofo (*PNAS* 1996 93 (15) 7436-7439) showed that materials with a certain electron structure have reduced electron thermal conductivity. Based on their analysis one can demonstrate that if the electron density of states in the material is close to the delta-function, the electronic thermal conductivity drops to zero. By taking the following equation $\lambda_E = \lambda_0 - T\sigma S^2$, where λ_0 is the electronic thermal conductivity when the electrochemical potential gradient inside the sample is zero, as a starting point. As next step the transport coefficients are written as following

$$\sigma = \sigma_0 I_0,$$

$$\sigma S = \left(\frac{k}{e}\right)\sigma_0 I_1$$

$$\lambda_0 = \left(\frac{k}{e}\right)^2 \sigma_0 T I_2$$

where $\sigma_0 = e^2 / (\hbar a_0)$ and a_0 the Bohr radius. The dimensionless integrals I_n are defined as

$$I_n = \int_{-\infty}^{\infty} \frac{e^x}{\left(e^x + 1\right)^2} s(x) x^n dx,$$

where $s(x)$ is the dimensionless transport distribution function. The integrals I_n are the moments of the function

$$P(x) = D(x)s(x),\ D(x) = \frac{e^x}{\left(e^x + 1\right)^2},$$

where x is the energy of carriers. By substituting the previous formulas for the transport coefficient to the equation for λ_E we get the following equation

$$\lambda_E = \left(\frac{k}{e}\right)^2 \sigma_0 T \left(I_2 - \frac{I_1^2}{I_0}\right).$$

From the previous equation we see that λ_E to be zero the bracketed term containing I_n

terms have to be zero. Now if we assume that

$$s(x) = f(x)\delta(x-b),,$$

where δ is the Dirac delta function, I_n terms get the following expressions

$$I_0 = D(b)f(b),$$

$$I_1 = D(b)f(b)b,$$

$$I_2 = D(b)f(b)b^2.$$

By substituting these expressions to the equation for λ_E, we see that it goes to zero. Therefore, $P(x)$ has to be delta function.

Equations

In an isotropic medium the thermal conductivity is the parameter k in the Fourier expression for the heat flux

$$\vec{q} = -k\vec{\nabla}T$$

where $\vec{q}$ is the heat flux (amount of heat flowing per second and per unit area) and $\vec{\nabla}T$ the temperature gradient. The sign in the expression is chosen so that always $k > 0$ as heat always flows from a high temperature to a low temperature. This is a direct consequence of the second law of thermodynamics.

In the one-dimensional case $q = H/A$ with H the amount of heat flowing per second through a surface with area A and the temperature gradient is $\mathrm{d}T/\mathrm{d}x$ so

$$H = -kA\frac{\mathrm{d}T}{\mathrm{d}x}.$$

In case of a thermally-insulated bar (except at the ends) in the steady state H is constant. If A is constant as well the expression can be integrated with the result

$$HL \quad A\int \quad k(T)\mathrm{d}T$$

where T_H and T_L are the temperatures at the hot end and the cold end respectively, and L is the length of the bar. It is convenient to introduce the thermal-conductivity integral

$$I_k(T) = \int_0^T k(T')\mathrm{d}T'.$$

The heat flow rate is then given by

$$H = \frac{A}{L}[I_k(T_H) - I_k(T_L)].$$

If the temperature difference is small k can be taken as constant. In that case

$$H = kA\frac{T_H - T_L}{L}.$$

Simple Kinetic Picture

In this section we will motivate an expression for the thermal conductivity in terms of microscopic parameters.

Consider a gas of particles of negligible volume governed hard-core interactions and within a vertical temperature gradient. The upper side is hot and the lower side cold. There is a downward energy flow because the gas atoms, going down, have a higher energy than the atoms going up. The net flow of energy per second is the heat flow H, which is proportional to the number of particles that cross the area A per second. In fact, H should also be proportional to the particle density n, the mean particle velocity v, the amount of energy transported per particle so with the heat capacity per particle c and some characteristic temperature difference ΔT. So far, in our model,

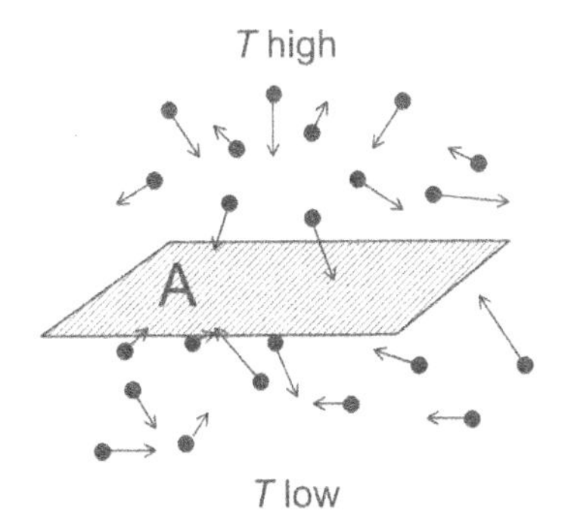

Gas atoms moving randomly through a surface.

$$H \propto nvcA\Delta T.$$

The unit of H is J/s and of the right-hand side it is (particle/m³)•(m/s)•(J/(K•particle))•(m²)•(K) = J/s, so this is already of the right dimension. Only a numerical factor is missing. For ΔT we take the temperature difference of the gas between two collisions $\Delta T = l\frac{dT}{dz}$ where l is the mean free path.

Detailed kinetic calculations show that the numerical factor is -1/3, so, all in all,

$$H = -\frac{1}{3}nvcl\,A\frac{dT}{dz}.$$

Comparison with the one-dimension expression for the heat flow, given above, gives an expression for the factor k

$$k = \frac{1}{3} nvcl.$$

The particle density and the heat capacity per particle can be combined as the heat capacity per unit volume

so
$$k = \frac{1}{3} vl \frac{C_V}{V_m}$$

where C_V is the molar heat capacity at constant volume and V_m the molar volume.

For the hard-core gas the mean free path is given by $l \propto \frac{1}{n\sigma}$ where σ is the collision cross section. So

$$k \propto \frac{c}{\sigma} v.$$

The heat capacity per particle c and the cross section σ both are temperature independent so the temperature dependence of k is determined by the T dependence of v. For a monatomic gas, with atomic mass M, v is given by $v = \sqrt{\frac{3RT}{M}}$. So

$$k \propto \sqrt{\frac{T}{M}}.$$

This expression also shows why gases with a low mass (hydrogen, helium) have a high thermal conductivity.

For metals at low temperatures the heat is carried mainly by the free electrons. In this case the mean velocity is the Fermi velocity which is temperature independent. The mean free path is determined by the impurities and the crystal imperfections which are temperature independent as well. So the only temperature-dependent quantity is the heat capacity c, which, in this case, is proportional to T. So

$$k = k_0 T \quad \text{(metal at low temperature)}$$

with k_0 a constant. For pure metals such as copper, silver, etc. l is large, so the thermal conductivity is high. At higher temperatures the mean free path is limited by the phonons, so the thermal conductivity tends to decrease with temperature. In alloys the density of the impurities is very high, so l and, consequently k, are small. Therefore, alloys, such as stainless steel, can be used for thermal insulation.

Thermal Insulation

Thermal insulation is the reduction of heat transfer (the transfer of thermal energy between objects of differing temperature) between objects in thermal contact or in range of radiative influence. Thermal insulation can be achieved with specially engineered methods or processes, as well as with suitable object shapes and materials.

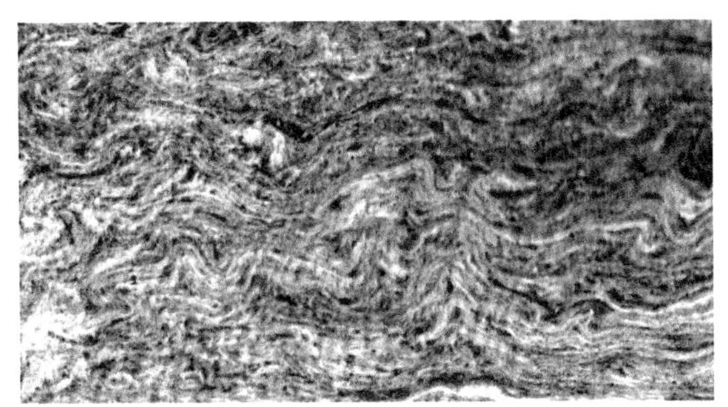

Mineral wool Insulation, 1600 dpi scan

Heat flow is an inevitable consequence of contact between objects of differing temperature. Thermal insulation provides a region of insulation in which thermal conduction is reduced or thermal radiation is reflected rather than absorbed by the lower-temperature body.

The insulating capability of a material is measured with thermal conductivity (k). Low thermal conductivity is equivalent to high insulating capability (R-value). In thermal engineering, other important properties of insulating materials are product density (ρ) and specific heat capacity (c).

Definition

Insulation

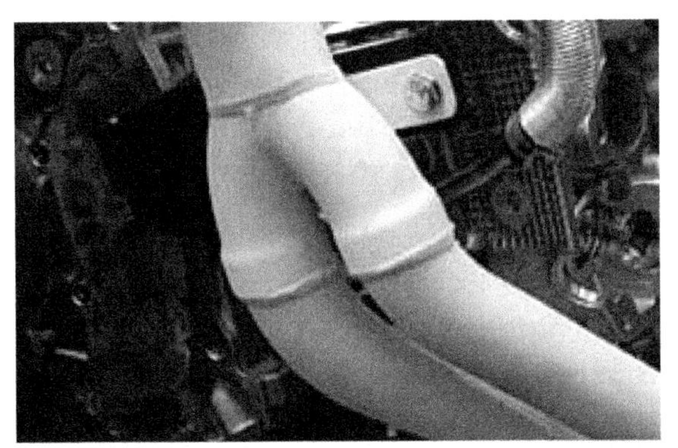

Car exhausts usually require some form of heat barrier, especially high performance exhausts where a ceramic coating is often applied

Solid materials chosen for insulation have a low thermal conductivity *k*, measured in watts-per-meter per kelvin ($W \cdot m^{-1} \cdot K^{-1}$). As the thickness of insulation is increased, the thermal resistance—or R-value—also increases.

For insulated cylinders, there is a complication - a *critical radius* beyond which extra insulation paradoxically increases heat transfer. The convective thermal resistance is inversely proportional to the surface area and therefore the radius of the cylinder, while the thermal resistance of a cylindrical shell (the insulation layer) depends on the ratio between outside and inside radius, not on the radius itself. If the outside radius of a cylinder is doubled by applying insulation, a fixed amount of conductive resistance (equal to $\ln(2)/(2\pi kL)$) is added. However, at the same time, the convective resistance has been halved. Because convective resistance tends to infinity when the radius approaches zero, at small enough radiuses the decrease in convective resistance will be larger than the added conductive resistance, resulting in lower total resistance. This implies that adding insulation actually increases the heat transfer, until a critical radius is reached, at which point the heat transfer is at maximum. Above this critical radius, added insulation decreases the heat transfer. For insulated cylinders, the critical radius is given by the equation

$$r_{critical} = \frac{k}{h}$$

This equation shows that the critical radius depends only on the heat transfer coefficient and the thermal conductivity of the insulation. If the radius of the uninsulated cylinder is larger than the critical radius for insulation, the addition of any amount of insulation will decrease the heat transfer.

Applications

Clothing and Natural Animal Insulation in Birds and Mammals

Gases possess poor thermal conduction properties compared to liquids and solids, and thus makes a good insulation material if they can be trapped. In order to further augment the effectiveness of a gas (such as air) it may be disrupted into small cells which cannot effectively transfer heat by natural convection. Convection involves a larger bulk flow of gas driven by buoyancy and temperature differences, and it does not work well in small cells where there is little density difference to drive it.

In order to accomplish gas cell formation in man-made thermal insulation, glass and polymer materials can be used to trap air in a foam-like structure. This principle is used industrially in building and piping insulation such as (glass wool), cellulose, rock wool, polystyrene foam (styrofoam), urethane foam, vermiculite, perlite, and cork. Trapping air is also the principle in all highly insulating clothing materials such as wool, down feathers and fleece.

The air-trapping property is also the insulation principle employed by homeothermic animals to stay warm, for example down feathers, and insulating hair such as natural sheep's wool. In both cases the primary insulating material is air, and the polymer used for trapping the air is natural keratin protein.

Buildings

Maintaining acceptable temperatures in buildings (by heating and cooling) uses a large proportion of global energy consumption. Building insulations also commonly use the principle of small trapped air-cells as explained above, e.g. fiberglass (specifically glass wool), cellulose, rock wool, polystyrene foam, urethane foam, vermiculite, perlite, cork, etc. For a period of time, Asbestos was also used, however, it caused health problems.

Common insulation applications in apartment building in Ontario, Canada.

When well insulated, a building:

- is energy-efficient, thus saving the owner money.
- provides more uniform temperatures throughout the space. There is less temperature gradient both vertically (between ankle height and head height) and horizontally from exterior walls, ceilings and windows to the interior walls, thus producing a more comfortable occupant environment when outside temperatures are extremely cold or hot.
- has minimal recurring expense. Unlike heating and cooling equipment, insulation is permanent and does not require maintenance, upkeep, or adjustment.
- lowers the carbon footprint of a building.

Many forms of thermal insulation also reduce noise and vibration, both coming from the outside and from other rooms inside a building, thus producing a more comfortable environment.

Window insulation film can be applied in weatherization applications to reduce incoming thermal radiation in summer and loss in winter.

In industry, energy has to be expended to raise, lower, or maintain the temperature of objects or process fluids. If these are not insulated, this increases the energy requirements of a process, and therefore the cost and environmental impact.

Mechanical Systems

Space heating and cooling systems distribute heat throughout buildings by means of pipe or ductwork. Insulating these pipes using pipe insulation reduces energy into unoccupied rooms and prevents condensation from occurring on cold and chilled pipework.

Thermal insulation applied to exhaust component by means of plasma spraying

Pipe insulation is also used on water supply pipework to help delay pipe freezing for an acceptable length of time.

Spacecraft

Launch and re-entry place severe mechanical stresses on spacecraft, so the strength of an insulator is critically important (as seen by the failure of insulating foam on the Space Shuttle Columbia). Re-entry through the atmosphere generates very high temperatures due to compression of the air at high speeds. Insulators must meet demanding physical properties beyond their thermal transfer retardant properties. E.g. reinforced carbon-carbon composite nose cone and silica fiber tiles of the Space Shuttle.

Thermal insulation on the Huygens probe

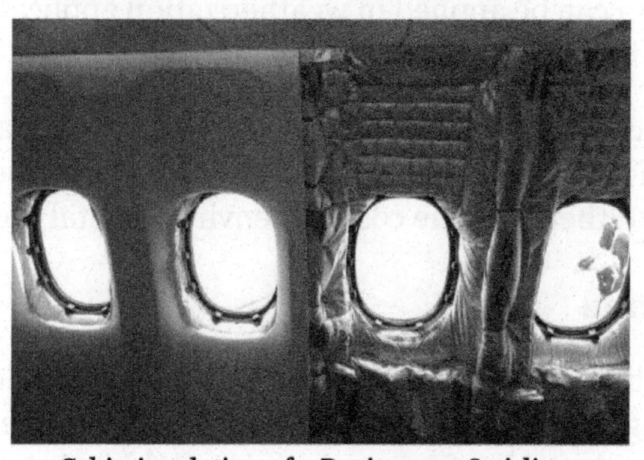

Cabin insulation of a Boeing 747-8 airliner

Automotive

Internal combustion engines produce a lot of heat during their combustion cycle. This can have a negative effect when it reaches various heat-sensitive components such as sensors, batteries and starter motors. As a result, thermal insulation is necessary to prevent the heat from the exhaust reaching these components.

High performance cars often use thermal insulation as a means to increase engine performance.

Factors Influencing Performance

Insulation performance is influenced by many factors the most prominent of which include:

- Thermal conductivity ("k" or "λ" value)
- Surface emissivity ("ε" value)
- Insulation thickness
- Density
- Specific heat capacity
- Thermal bridging

It is important to note that the factors influencing performance may vary over time as material ages or environmental conditions change.

Calculating Requirements

Industry standards are often rules of thumb, developed over many years, that offset many conflicting goals: what people will pay for, manufacturing cost, local climate, traditional building practices, and varying standards of comfort. Both heat transfer and layer analysis may be performed in large industrial applications, but in household situations (appliances and building insulation), air tightness is the key in reducing heat transfer due to air leakage (forced or natural convection). Once air tightness is achieved, it has often been sufficient to choose the thickness of the insulating layer based on rules of thumb. Diminishing returns are achieved with each successive doubling of the insulating layer. It can be shown that for some systems, there is a minimum insulation thickness required for an improvement to be realized.

Thermal Equilibrium

Two physical systems are in thermal equilibrium if no heat flows between them when they are connected by a path permeable to heat. Thermal equilibrium obeys the zeroth law of thermodynamics. A system is said to be in thermal equilibrium with itself if the temperature within the system is spatially and temporally uniform.

Systems in thermodynamic equilibrium are always in thermal equilibrium, but the converse is not always true. If the connection between the systems allows transfer of energy as heat but does not allow transfer of matter or transfer of energy as work, the two systems may reach thermal equilibrium without reaching thermodynamic equilibrium.

Two Varieties of Thermal Equilibrium

Relation of Thermal Equilibrium Between Two Thermally Connected Bodies

The relation of thermal equilibrium is an instance of a contact equilibrium between two bodies, which means that it refers to transfer through a selectively permeable partition, the contact path. For the relation of thermal equilibrium, the contact path is permeable only to heat; it does not permit the passage of matter or work; it is called a diathermal connection. According to Lieb and Yngvason, the essential meaning of the relation of thermal equilibrium includes that it is reflexive and symmetric. It is not included in the essential meaning whether it is or is not transitive. After discussing the semantics of the definition, they postulate a substantial physical axiom, that they call the "zeroth law of thermodynamics", that thermal equilibrium is a transitive relation. They comment that the equivalence classes of systems so established are called isotherms.

Internal Thermal Equilibrium of an Isolated Body

Thermal equilibrium of a body in itself refers to the body when it is isolated. The background is that no heat enters or leaves it, and that it is allowed unlimited time to settle under its own intrinsic characteristics. When it is completely settled, so that macroscopic change is no longer detectable, it is in its own thermal equilibrium. It is not implied that it is necessarily in other kinds of internal equilibrium. For example, it is possible that a body might reach internal thermal equilibrium but not be in internal chemical equilibrium; glass is an example.

One may imagine an isolated system, initially not in its own state of internal thermal equilibrium. It could be subjected to a fictive thermodynamic operation of partition into two subsystems separated by nothing, no wall. One could then consider the possibility of transfers of energy as heat between the two subsystems. A long time after the fictive partition operation, the two subsystems will reach a practically stationary state, and so be in the relation of thermal equilibrium with each other. Such an adventure could be conducted in indefinitely many ways, with different fictive partitions. All of them will result in subsystems that could be shown to be in thermal equilibrium with each other, testing subsystems from different partitions. For this reason, an isolated system, initially not its own state of internal thermal equilibrium, but left for a long time, practically always will reach a final state which may be regarded as one of internal thermal equilibrium. Such a final state is one of spatial uniformity or homogeneity of temperature. The existence of such states is a basic postulate of classical thermodynamics. This postulate is sometimes, but not often, called the minus first law of thermodynamics.

Thermal Contact

Heat can flow into or out of a closed system by way of thermal conduction or of thermal radiation to or from a thermal reservoir, and when this process is effecting net transfer of heat, the system is not in thermal equilibrium. While the transfer of energy as heat continues, the system's temperature can be changing.

Bodies Prepared with Separately Uniform Temperatures, then Put Into Purely Thermal Communication with Each Other

If bodies are prepared with separately macroscopically stationary states, and are then put into purely thermal connection with each other, by conductive or radiative pathways, they will be in thermal equilibrium with each other just when the connection is followed by no change in either body. But if initially they are not in a relation of thermal equilibrium, heat will flow from the hotter to the colder, by whatever pathway, conductive or radiative, is available, and this flow will continue until thermal equilibrium is reached and then they will have the same temperature.

One form of thermal equilibrium is radiative exchange equilibrium. Two bodies, each with its own uniform temperature, in solely radiative connection, no matter how far apart, or what partially obstructive, reflective, or refractive, obstacles lie in their path of radiative exchange, not moving relative to one another, will exchange thermal radiation, in net the hotter transferring energy to the cooler, and will exchange equal and opposite amounts just when they are at the same temperature. In this situation, Kirchhoff's law of equality of radiative emissivity and absorptivity and the Helmholtz reciprocity principle are in play.

Change of Internal State of an Isolated System

If an initially isolated physical system, without internal walls that establish adiabatically isolated subsystems, is left long enough, it will usually reach a state of thermal equilibrium in itself, in which its temperature will be uniform throughout, but not necessarily a state of thermodynamic equilibrium, if there is some structural barrier that can prevent some possible processes in the system from reaching equilibrium; glass is an example. Classical thermodynamics in general considers idealized systems that have reached internal equilibrium, and idealized transfers of matter and energy between them.

An isolated physical system may be inhomogeneous, or may be composed of several subsystems separated from each other by walls. If an initially inhomogeneous physical system, without internal walls, is isolated by a thermodynamic operation, it will in general over time change its internal state. Or if it is composed of several subsystems separated from each other by walls, it may change its state after a thermodynamic operation that changes its walls. Such changes may include change of temperature or spatial distribution of temperature, by changing the state of constituent materials. A rod of iron, initially prepared to be hot at one end and cold at the other, when isolated, will change so that its temperature becomes uniform all along its length; during the process, the rod is not in thermal equilibrium until its temperature is uniform. In a system prepared as a block of ice floating in a bath of hot water, and then isolated, the ice can melt; during the melting, the system is not in thermal equilibrium; but eventually its temperature will become uniform; the block of ice will not re-form. A system prepared as a mixture of petrol vapour and air can be ignited by a spark and produce carbon dioxide and water; if this happens in an isolated system, it will increase the temperature of the system, and during the increase, the system is not in thermal equilibrium; but eventually the system will settle to a uniform temperature.

Such changes in isolated systems are irreversible in the sense that while such a change will occur spontaneously whenever the system is prepared in the same way, the reverse change will practically never occur spontaneously within the isolated system; this is a large part of the content of the second law of thermodynamics. Truly perfectly isolated systems do not occur in nature, and always are artificially prepared.

In a Gravitational Field

One may consider a system contained in a very tall adiabatically isolating vessel with rigid walls initially containing a thermally heterogeneous distribution of material, left for a long time under the influence of a steady gravitational field, along its tall dimension, due to an outside body such as the earth. It will settle to a state of uniform temperature throughout, though not of uniform pressure or density, and perhaps containing several phases. It is then in internal thermal equilibrium and even in thermodynamic equilibrium. This means that all local parts of the system are in mutual radiative exchange equilibrium. This means that the temperature of the system is spatially uniform. This is so in all cases, including those of non-uniform external force fields. For an externally imposed gravitational field, this may be proved in macroscopic thermodynamic terms, by the calculus of variations, using the method of Langrangian multipliers. Considerations of kinetic theory or statistical mechanics also support this statement.

Distinctions Between Thermal and Thermodynamic Equilibria

There is an important distinction between thermal and thermodynamic equilibrium. According to Münster (1970), in states of thermodynamic equilibrium, the state variables of a system do not change at a measurable rate. Moreover, "The proviso 'at a measurable rate' implies that we can consider an equilibrium only with respect to specified processes and defined experimental conditions." Also, a state of thermodynamic equilibrium can be described by fewer macroscopic variables than any other state of a given body of matter. A single isolated body can start in a state which is not one of thermodynamic equilibrium, and can change till thermodynamic equilibrium is reached. Thermal equilibrium is a relation between two bodies or closed systems, in which transfers are allowed only of energy and take place through a partition permeable to heat, and in which the transfers have proceeded till the states of the bodies cease to change.

An explicit distinction between 'thermal equilibrium' and 'thermodynamic equilibrium' is made by C.J. Adkins. He allows that two systems might be allowed to exchange heat but be constrained from exchanging work; they will naturally exchange heat till they have equal temperatures, and reach thermal equilibrium, but in general will not be in thermodynamic equilibrium. They can reach thermodynamic equilibrium when they are allowed also to exchange work.

Another explicit distinction between 'thermal equilibrium' and 'thermodynamic equilibrium' is made by B. C. Eu. He considers two systems in thermal contact, one a thermometer, the other a system in which several irreversible processes are occurring. He considers the case in which, over the time scale of interest, it happens that both the thermometer reading and the irreversible processes are steady. Then there is thermal equilibrium without thermodynamic equilibrium. Eu proposes consequently that the zeroth law of thermodynamics can be considered to apply even when thermodynamic equilibrium is not present; also he proposes that if changes are occurring so fast that a steady temperature cannot be defined, then "it is no longer possible to describe the process by means of a thermodynamic formalism. In other words, thermodynamics has no meaning for such a process."

Interfacial Thermal Resistance

Interfacial thermal resistance, also known as thermal boundary resistance, or Kapitza resistance, is a measure of an interface's resistance to thermal flow. This thermal resistance differs from contact resistance as it exists even at atomically perfect interfaces. Due to the differences in electronic and vibrational properties in different materials, when an energy carrier (phonon or electron, depending on the material) attempts to traverse the interface, it will scatter at the interface. The probability of transmission after scattering will depend on the available energy states on side 1 and side 2 of the interface.

Assuming a constant thermal flux is applied across an interface, this interfacial thermal resistance will lead to a finite temperature discontinuity at the interface. From an extension of Fourier's law, we can write

$$Q = \frac{\Delta T}{R} = G\Delta T$$

where Q is the applied flux, ΔT is the observed temperature drop, R is the thermal boundary resistance, and G is its inverse, or thermal boundary conductance.

Understanding the thermal resistance at the interface between two materials is of primary significance in the study of its thermal properties. Interfaces often contribute significantly to the observed properties of the materials. This is even more critical for nanoscale systems where interfaces could significantly affect the properties relative to bulk materials.

Low thermal resistance at interfaces is technologically important for applications where very high heat dissipation is necessary. This is of particular concern to the development of microelectronic semiconductor devices as defined by the International Technology Roadmap for Semiconductors in 2004 where an 8 nm feature size device is projected to generate up to 100000 W/cm^2 and would need efficient heat dissipation of an anticipated die level heat flux of 1000 W/cm^2 which is an order of magnitude higher than current devices. On the other hand, applications requiring good thermal isolation such as jet engine turbines would benefit from interfaces with high thermal resistance. This would also require material interfaces which are stable at very high temperature. Examples are metal-ceramic composites which are currently used for these applications. High thermal resistance can also be achieved with multilayer systems.

As stated above, thermal boundary resistance is due to carrier scattering at an interface. The type of carrier scattered will depend on the materials governing the interfaces. For example, at a metal-metal interface, electron scattering effects will dominate thermal boundary resistance, as electrons are the primary thermal energy carriers in metals.

Two widely used predictive models are the acoustic mismatch model (AMM) and the diffuse mismatch model (DMM). The AMM assumes a geometrically perfect interface and phonon transport across it is entirely elastic, treating phonons as waves in a continuum. On the other hand, the DMM assumes scattering at the interface is diffusive, which is accurate for interfaces with characteristic roughness at elevated temperatures.

Molecular dynamics (MD) simulations are a powerful tool to investigate interfacial thermal resistance. Recent MD studies have demonstrated that the solid-liquid interfacial thermal resistance is reduced on nanostructured solid surfaces by enhancing the solid-liquid interaction energy per unit area, and reducing the difference in vibrational density of states between solid and liquid.

Theoretical Models

There are two primary models that are used to understand the thermal resistance of interfaces, the acoustic mismatch and diffuse mismatch models (AMM and DMM respectively). Both models are based only on phonon transport, ignoring electrical contributions. Thus it should apply for interfaces where at least one of the materials is electrically insulating. For both models the interface is assumed to behave exactly as the bulk on either side of the interface (e.g. bulk phonon dispersions, velocities, etc.). The thermal resistance then results from the transfer of phonons across the interface. Energy is transferred when higher energy phonons which exist in higher density in the hotter material propagate to the cooler materials, which in turn transmits lower energy phonons, creating a net energy flux.

A crucial factor in determining the thermal resistance at an interface is the overlap of phonon states. Given two materials, A and B, if material A has a low population (or no population) of phonons with certain k value, there will be very few phonons of that wavevector to propagate from A to B. Further, due to the detailed balance, very few phonons of that wavevector will propagate the opposite direction, from B to A, even if material B has a large population of phonons with that wavevector. Thus as the overlap between phonon dispersions is small, there are less modes to allow for heat transfer in the material, giving at a high thermal interfacial resistance relative to materials with a high degree of overlap. Both AMM and DMM reflect this principle, but differ in the conditions they require for propagation across the interface. Neither model is universally effective for predicting the thermal interface resistance (with the exception of very low temperature), but rather for most materials they act as upper and lower limits for real behavior.

Both models differ greatly in their treatment of scattering at the interface. In AMM the interface is assumed to be perfect, resulting in no scattering, thus phonons propagate elastically across the interface. The wavevectors that propagate across the interface are determined by conservation of momentum. In DMM, the opposite extreme is assumed, a perfectly scattering interface. In this case the wavevectors that propagate across the interface are random and independent of incident phonons on the interface. For both models the detailed balance must still be obeyed.

For both models some basic equations apply. The flux of energy from one material to the other is just:

$$Q_{1,2} = \sum_{k} n(k, T_1) E(k) \alpha(k, T_1, T_2)$$

where n is the number of phonons at a given wavevector and momentum, E is the energy, and α is the probability of transmission across the interface. The net flux is thus the difference of the energy fluxes:

$$Q_{net} = Q_{1,2} - Q_{2,1}$$

Since both fluxes are dependent on T_1 and T_2, the relationship between the flux and the temperature difference can be used to determine the thermal interface resistance based on:

$$R_{th} = \frac{\Delta T}{Q / A}$$

where A is the area of the interface. These basic equations form the basis for both models. n is determined based on the Debye model and Bose–Einstein statistics. Energy is given simply by:

$$E = \hbar\, \omega(k)\, v$$

where v is the speed of sound in the material. The main difference between the two models is the transmission probability, whose determination is more complicated. In each case it is determined by the basic assumptions that form the respective models. The assumption of elastic scattering makes it more difficult for phonons to transmit across the interface, resulting in lower probabilities. As a result, the acoustic mismatch model typically represents an upper limit for thermal interface resistance, while the diffuse mismatch model represents the lower limit.

Examples

Liquid Helium Interfaces

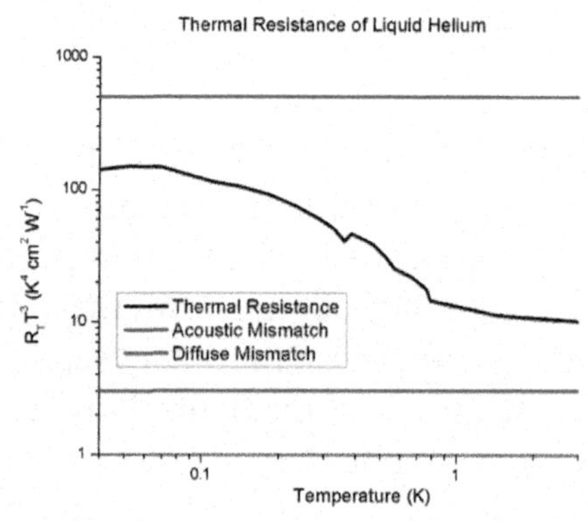

Typical Interfacial Resistance of Liquid Helium with metals. Resistance has been multiplied by T3 to remove the expected T−3 dependence. Adapted from .

The presence of thermal interface resistance, corresponding to a discontinuous temperature across an interface was first proposed from studies of liquid helium in 1936. While this idea was first proposed in 1936, it wasn't until 1941 when Pyotr Kapitsa (Peter Kapitza) carried out the first systematic study of thermal interface behavior in liquid helium. The first major model for heat transfer at interfaces was the acoustic mismatch model which predicted a T^{-3} temperature dependence on the interfacial resistance, but this failed to properly model the thermal conductance of helium interfaces by as much as two orders of magnitude. Another surprising behavior of the thermal resistance was observed in the pressure dependence. Since the speed of sound is a strong function of temperature in liquid helium, the acoustic mismatch model predicts a strong pressure dependence of the interfacial resistance. Studies around 1960 surprisingly showed that the interfacial resistance was nearly independent of pressure, suggesting that other mechanisms were dominant

The acoustic mismatch theory predicted a very high thermal resistance (low thermal conductance) at solid-helium interfaces. This was potentially disastrous to researchers working at ultra-cold temperatures because it greatly impedes cooling rates at low temperatures. Fortunately such a large thermal resistance was not observed due to many mechanisms which promoted phonon transport. In liquid helium, Van der Waals forces actually work to solidify the first few monolayers against a solid. This boundary layer functions much like an anti-reflection coating in optics, so that phonons which would typically be reflected from the interface actually would transmit across the interface. This also helps to understand the pressure independence of the thermal conductance. The final dominant mechanism to anomalously low thermal resistance of liquid helium interfaces is the effect of surface roughness, which is not accounted for in the acoustic mismatch model. For a more detailed theoretical model of this aspect see the paper by A. Khater and J. Szeftel. Like electromagnetic waves which produce surface plasmons on rough surfaces, phonons can also induce surface waves. When these waves eventually scatter, they provide another mechanism for heat to transfer across the interface. Similarly, phonons are also capable of producing evanescent waves in a total internal reflection geometry. As a result, when these waves are scattered in the solid, additional heat is transferred from the helium beyond the prediction of the acoustic mismatch theory.

Notable Room Temperature Thermal Conductance

In general there are two types of heat carriers in materials: phonons and electrons. The free electron gas found in metals is a very good conductor of heat and dominates thermal conductivity. All materials though exhibit heat transfer by phonon transport so heat flows even in dielectric materials such as silica. Interfacial thermal conductance is a measure of how efficiently heat carriers flow from one material to another. The lowest room temperature thermal conductance measurement to date is the Bi/Hydrogen-terminated diamond with a thermal conductance of 8.5 MW m^{-2} K^{-1}. As a metal, bismuth contains many electrons which serve as the primary heat carriers. Diamond on the other hand is a very good electrical insulator (although it has a very high thermal conductivity) and so electron transport between the materials is nil. Further, these materials have very different lattice parameters so phonons do not efficiently couple across the interface. Finally, the Debye temperature between the materials is significantly different. As a result, bismuth, which has a low Debye temperature, has many phonons at low frequencies. Diamond on the other hand has a very high Debye temperature and most of its heat-carrying phonons are at frequencies much higher than are present in bismuth.

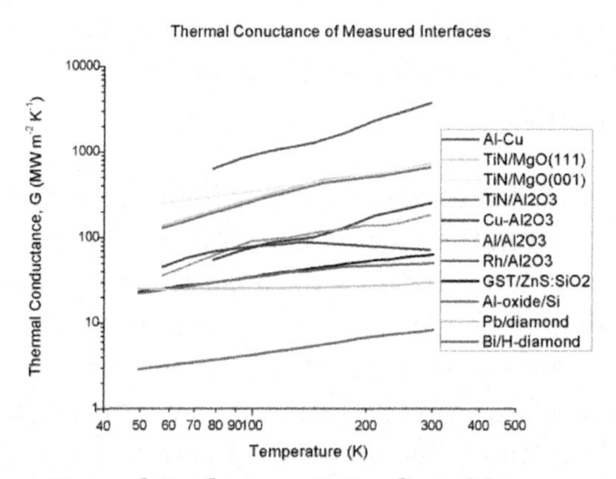

Thermal Conductance Data adapted from,

Increasing in thermal conductance, most phonon mediated interfaces (dielectric-dielectric and metal-dielectric) have thermal conductances between 80 and 300 MW m^{-2} K^{-1}. The largest phonon mediated thermal conductance measured to date is between TiN (Titanium Nitride) and MgO. These systems have very similar lattice structures and Debye temperatures. While there are no free electrons to enhance the thermal conductance of the interface, the similar physical properties of the two crystals facilitate a very efficient phonon transmission between the two materials.

At the highest end of the spectrum, one of the highest thermal conductances *measured* is between aluminum and copper. At room temperature, the Al-Cu interface has a conductance of 4 GW m^{-2} K^{-1}. The high thermal conductance of the interface should not be unexpected given the high electrical conductivity of both materials.

Interfacial Resistance in Carbon Nanotubes

The superior thermal conductivity of Carbon nanotubes makes it an excellent candidate for making composite materials. But interfacial resistance impacts the effective thermal conductivity. This area is not well studied and only a few studies have been done to understand the basic mechanism of this resistance.

Waste Heat

Waste heat is by necessity produced both by machines that do work and in other processes that use energy, for example in a refrigerator warming the room air or a combustion engine releasing heat into the environment. The need for many systems to reject heat as a by-product of their operation is fundamental to the laws of thermodynamics. Waste heat has lower utility (or in thermodynamics lexicon a lower exergy or higher entropy) than the original energy source. Sources of waste heat include all manner of human activities, natural systems, and all organisms. Rejection of unneeded cold(as from a heat pump) is also a form of waste heat (i.e. the medium has heat, but at a lower temperature than is considered warm).

Air conditioning units use electricity which ends up as heat

Instead of being "wasted" by release into the ambient environment, sometimes waste heat (or cold) can be utilized by another process, or a portion of heat that would otherwise be wasted can be reused in the same process if make-up heat is added to the system (as with heat recovery ventilation in a building).

Thermal energy storage, which includes technologies both for short- and long-term retention of heat or cold, can create or improve the utility of waste heat (or cold). One example is waste heat from air conditioning machinery stored in a buffer tank to aid in night time heating. Another is seasonal thermal energy storage (STES) at a foundry in Sweden. The heat is stored in the bedrock surrounding a cluster of heat exchanger equipped boreholes, and is used for space heating in an adjacent factory as needed, even months later. An example of using STES to utilize natural waste heat is the Drake Landing Solar Community in Alberta, Canada, which, by using a cluster of boreholes in bedrock for interseasonal heat storage, obtains 97 percent of its year-round heat from solar thermal collectors on the garage roofs. Another STES application is storing winter cold underground, for summer air conditioning.

On a biological scale, all organisms reject waste heat as part of their metabolic processes, and will die if the ambient temperature is too high to allow this.

Anthropogenic waste heat is thought by some to contribute to the urban heat island effect. The biggest point sources of waste heat originate from machines (such as electrical generators or industrial processes, such as steel or glass production) and heat loss through building envelopes. The burning of transport fuels is a major contribution to waste heat.

Conversion of Energy

Machines converting energy contained in fuels to mechanical work or electric energy produce heat as a by-product.

Sources

In the majority of energy applications, energy is required in multiple forms. These energy forms typically include some combination of: heating, ventilation, and air conditioning, mechanical energy and electric power. Often, these additional forms of energy are produced by a heat engine, running on a source of high-temperature heat. A heat engine can never have perfect efficiency, according to the second law of thermodynamics, therefore a heat engine will always produce a surplus of low-temperature heat. This is commonly referred to as waste heat or "secondary heat", or "low-grade heat". This heat is useful for the majority of heating applications, however, it is sometimes not practical to transport heat energy over long distances, unlike electricity or fuel energy. The largest proportions of total waste heat are from power stations and vehicle engines.The largest single sources are power stations and industrial plants such as oil refineries and steelmaking plants.

Power Generation

The electrical efficiency of thermal power plants is defined as the ratio between the input and output energy. It is typically only 30%. The images show cooling towers which allow power stations to maintain the low side of the temperature difference essential for conversion of heat differences to other forms of energy. Discarded or "Waste" heat that is lost to the environment may instead be used to advantage.

Coal-fired power station that transform chemical energy into 36%-48% electricity and remaining 52%-64% to waste heat

Industrial Processes

Industrial processes, such as oil refining, steel making or glass making are major sources of waste heat.

Electronics

Although small in terms of power, the disposal of waste heat from microchips and other electronic components, represents a significant engineering challenge. This necessitates the use of fans, heatsinks, etc. to dispose of the heat.

Biological

Animals, including humans, create heat as a result of metabolism. In warm conditions, this heat exceeds a level required for homeostasis in warm-blooded animals, and is disposed of by various thermoregulation methods such as sweating and panting. Fiala *et al.* modelled human thermoregulation.

Cooling towers evaporating water at Ratcliffe-on-Soar Power Station, United Kingdom.

Disposal

Low temperature heat contains very little capacity to do work (Exergy), so the heat is qualified as waste heat and rejected to the environment. Economically most convenient is the rejection of such heat to water from a sea, lake or river. If sufficient cooling water is not available, the plant has to be equipped with a cooling tower to reject the waste heat into the atmosphere. In some cases it is possible to use waste heat, for instance in heating homes by cogeneration. However, by slowing the release of the waste heat, these systems always entail a reduction of efficiency for the primary user of the heat energy.

Uses

Cogeneration and Trigeneration

Waste of the by-product heat is reduced if a cogeneration system is used, also known as a Combined Heat and Power (CHP) system. Limitations to the use of by-product heat arise primarily from the engineering cost/efficiency challenges in effectively utilizing small temperature differences to generate other forms of energy. Applications utilizing waste heat include swimming pool heating, paper mills. In some cases cooling can also be produced by the use of absorption refrigerators for example, in this case it's called trigeneration or CCHP (combined cooling, heat and power).

Pre-heating

Waste heat can be forced to heat incoming fluids and objects before being highly heated. For instance outgoing water can give its waste heat to incoming water in a heat exchanger before heating in homes or power plants.

Electrification of Waste Heat

There are many different approaches to transfer thermal energy to electricity, and the technologies to do so have existed for several decades. The organic Rankine cycle, offered by companies such as Ormat, is a very known approach, whereby an organic substance is used as working medium instead of water. The benefit is that this process can utilize lower temperatures for the production of electricity than the regular water steam cycle. An example of use of the steam Rankine cycle is the Cyclone Waste Heat Engine. Another established approach is by using a thermoelectric, such as those offered by Alphabet Energy, where a change in temperature across a semiconductor material creates a voltage through a phenomenon known as the Seebeck effect. A related approach is the use of thermogalvanic cells, where a temperature difference gives rise to an electric current in an electrochemical cell.

Greenhouses

Waste heat (along with carbon dioxide from combustion) can be used to provide heat for greenhouses particularly in colder climates.

Anthropogenic Heat

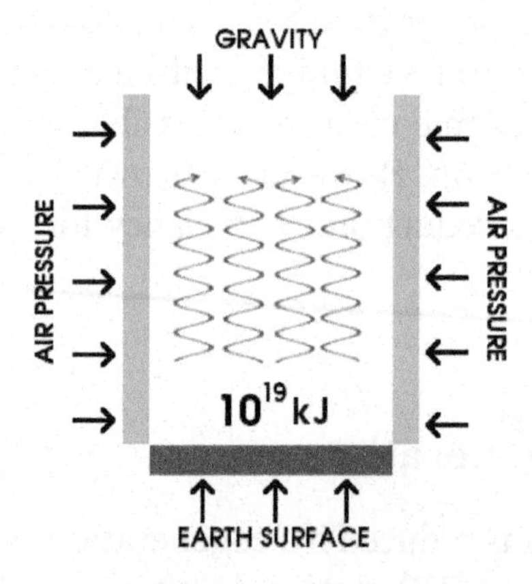

Anthropogenic heat

Anthropogenic heat is heat generated by humans and human activity. The American Meteorological Society defines it as "Heat released to the atmosphere as a result of human activities, often involving combustion of fuels. Sources include industrial plants, space heating and cooling, human metabolism, and vehicle exhausts. In cities this source typically contributes 15–50 W/m^2 to the local heat balance, and several hundred W/m^2 in the center of large cities in cold climates and industrial areas."

Estimates of anthropogenic heat generation can be made by totaling all the energy used for heating and cooling, running appliances, transportation, and industrial processes, plus that directly emitted by human metabolism.

Environmental Impact

Anthropogenic heat is a small influence on rural temperatures, and becomes more significant in dense urban areas. It is one contributor to urban heat islands. Other human-caused effects (such as changes to albedo, or loss of evaporative cooling) that might contribute to urban heat islands are not considered to be anthropogenic heat by this definition.

Anthropogenic heat is a much smaller contributor to global warming than are greenhouse gases. In 2005, although anthropogenic waste heat flux was significantly high in certain urban areas (and can be high regionally. For example, waste heat flux was +0.39 and +0.68 W/m^2 for the continental United States and western Europe, respectively) globally it accounted for only 1% of the energy flux created by anthropogenic greenhouse gases. Global forcing from waste heat was 0.028 W/m^2 in 2005. This statistic is predicted to rise as urban areas become more widespread.

Although waste heat has been shown to have influence on regional climates, climate forcing from waste heat is not normally calculated in state-of-the-art global climate simulations. Equilibrium climate experiments show statistically significant continental-scale surface warming (0.4–0.9 °C) produced by one 2100 AHF scenario, but not by current or 2040 estimates. Simple global-scale estimates with different growth rates of anthropogenic heat that have been actualized recently show noticeable contributions to global warming, in the following centuries. For example, a 2% p.a. growth rate of waste heat resulted in a 3 degree increase as a lower limit for the year 2300. Meanwhile, this has been confirmed by more refined model calculations.

Heat Flux

Heat flux or thermal flux is the rate of heat energy transfer through a given surface per unit time. The SI derived unit of *heat rate* is joule per second, or watt. Heat flux density is the heat rate per unit area. In SI units, heat flux density is measured in [W/m^2]. Heat rate is a scalar quantity, while heat flux is a vectorial quantity. To define the heat flux at a certain point in space, one takes the limiting case where the size of the surface becomes infinitesimally small.

Heat flux is often denoted $\vec{\phi}_q$, the subscript q specifying *heat* rate, as opposed to *mass* or *momentum* rate. Fourier's law is an important application of these concepts.

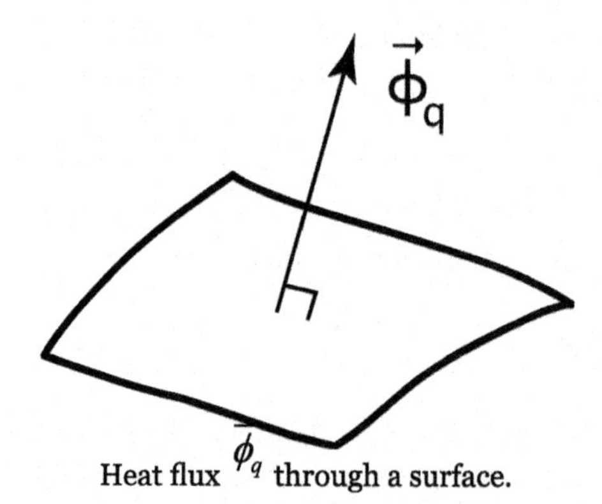

Heat flux $\vec{\phi}_q$ through a surface.

Fourier's Law

For most solids in usual conditions, heat is transported mainly by conduction and the heat flux is adequately described by Fourier's law.

Fourier's Law in One Dimension

The heat flux associated with a temperature profile $T(x)$ in a material of thermal conductivity k is given by

$$\phi_q = -k\frac{dT(x)}{dx}$$

The negative sign shows that heat flux moves from higher temperature regions to lower temperature regions.

Multi-dimensional Extension

The multi-dimensional case is similar, the heat flux goes "down" the temperature gradient hence the negative sign:

$$\vec{q} = -k\nabla T$$

where ∇ is the gradient operator.

Measuring Heat Flux

The measurement of heat flux is most often done by measuring a temperature difference over a piece of material with known thermal conductivity. This method is analogous to a standard way to measure an electric current, where one measures the voltage drop over a known resistor.

Relevance to Science and Engineering

One of the tools in a scientist's or engineer's toolbox is the energy balance. Such a balance can be set up for any physical system, from chemical reactors to living organisms, and generally takes the following form

$$\frac{\partial E_{\text{in}}}{\partial t} - \frac{\partial E_{\text{out}}}{\partial t} - \frac{\partial E_{\text{accumulated}}}{\partial t} = 0$$

where the three $\frac{\partial E}{\partial t}$ terms stand for the time rate of change of respectively the total amount of in

coming energy, the total amount of outgoing energy and the total amount of accumulated energy.

Now, if the only way the system exchanges energy with its surroundings is through heat transfer, the heat rate can be used to calculate the energy balance, since

where we have integrated the heat flux density $\vec{\phi}_q$ over the surface S of the system.

In real-world applications one cannot know the exact heat flux at every point on the surface, but approximation schemes can be used to calculate the integral, for example Monte Carlo integration.

Heat Transfer Coefficient

The heat transfer coefficient or film coefficient, or film effectiveness, in thermodynamics and in mechanics is the proportionality constant between the heat flux and the thermodynamic driving force for the flow of heat (i.e., the temperature difference, ΔT):

$$h = \frac{q}{\Delta T}$$

where:

q: amount of heat transferred (heat flux), W/m^2 i.e., thermal power per unit area, $q = d\dot{Q}/dA$

h: heat transfer coefficient, W/(m^2•K)

ΔT: difference in temperature between the solid surface and surrounding fluid area, K.

It is used in calculating the heat transfer, typically by convection or phase transition between a fluid and a solid. The heat transfer coefficient has SI units in watts per squared meter kelvin: W/(m^{2}K).

The heat transfer coefficient is the reciprocal of thermal insulance. This is used for building materials (R-value) and for clothing insulation.

There are numerous methods for calculating the heat transfer coefficient in different heat transfer modes, different fluids, flow regimes, and under different thermohydraulic conditions. Often it can be estimated by dividing the thermal conductivity of the convection fluid by a length scale. The heat transfer coefficient is often calculated from the Nusselt number (a dimensionless number). There are also online calculators available specifically for heat transfer fluid applications. Experimental assessment of the heat transfer coefficient poses some challenges especially when small fluxes are to be measured (e.g. $< 0.2W/cm^2$).

Composition

A simple method for determining an overall heat transfer coefficient that is useful to find the heat transfer between simple elements such as walls in buildings or across heat exchangers is shown below. Note that this method only accounts for conduction within materials, it does not take into account heat transfer through methods such as radiation. The method is as follows:

$$1/(U \cdot A) = 1/(h_1 \cdot A_1) + dx_w/(k \cdot A) + 1/(h_2 \cdot A_2)$$

Where:

- U = the overall heat transfer coefficient (W/(m²•K))
- A = the contact area for each fluid side (m²) (with A_1 and A_2 expressing either surface)
- k = the thermal conductivity of the material (W/(m·K))
- h = the individual convection heat transfer coefficient for each fluid (W/(m²•K))
- dx_w = the wall thickness (m).

As the areas for each surface approach being equal the equation can be written as the transfer coefficient per unit area as shown below:

$$1/U=1/h_1+dx_w/k+1/h_2$$

or

$$U=1/(1/h_1+dx_w/k+1/h_2)$$

It is to be noted that often the value for dx_w is referred to as the difference of two radii where the inner and outer radii are used to define the thickness of a pipe carrying a fluid, however, this figure may also be considered as a wall thickness in a flat plate transfer mechanism or other common flat surfaces such as a wall in a building when the area difference between each edge of the transmission surface approaches zero.

In the walls of buildings the above formula can be used to derive the formula commonly used to calculate the heat through building components. Architects and engineers call the resulting values either the U-Value or the R-Value of a construction assembly like a wall. Each type of value (R or U) are related as the inverse of each other such that R-Value = 1/U-Value and both are more fully understood through the concept of an overall heat transfer coefficient described in lower section of this document.

Convective Heat Transfer Correlations

Although convective heat transfer can be derived analytically through dimensional analysis, exact analysis of the boundary layer, approximate integral analysis of the boundary layer and analogies between energy and momentum transfer, these analytic approaches may not offer practical solutions to all problems when there are no mathematical models applicable. Therefore, many correlations were developed by various authors to estimate the convective heat transfer coefficient in various cases including natural convection, forced convection for internal flow and forced convection for external flow. These empirical correlations are presented for their particular geometry and flow conditions. As the fluid properties are temperature dependent, they are evaluated at the film temperature T_f, which is the average of the surface T_s and the surrounding bulk temperature, T_∞.

$$T_f = \frac{T_s + T_\infty}{2}$$

External Flow, Vertical Plane

Recommendations by Churchill and Chu provide the following correlation for natural convection adjacent to a vertical plane, both for laminar and turbulent flow. k is the thermal conductivity of the fluid, L is the characteristic length with respect to the direction of gravity, Ra_L is the Rayleigh number with respect to this length and Pr is the Prandtl number.

$$h = \frac{k}{L}\left(0.825 + \frac{0.387\mathrm{Ra}_L^{1/6}}{\left(1 + (0.492/\mathrm{Pr})^{9/16}\right)^{8/27}}\right)^2 \qquad \mathrm{Ra}_L < 10^{12}$$

For laminar flows, the following correlation is slightly more accurate. It is observed that a transition from a laminar to a turbulent boundary occurs when Ra_L exceeds around 10^9.

$$h = \frac{k}{L}\left(0.68 + \frac{0.67\mathrm{Ra}_L^{1/4}}{\left(1 + (0.492/\mathrm{Pr})^{9/16}\right)^{4/9}}\right) \qquad 10^{-1} < \mathrm{Ra}_L < 10^9$$

External Flow, Vertical Cylinders

For cylinders with their axes vertical, the expressions for plane surfaces can be used provided the curvature effect is not too significant. This represents the limit where boundary layer thickness is small relative to cylinder diameter D. The correlations for vertical plane walls can be used when $\frac{D}{L} \geq \frac{35}{Gr_L^{\frac{1}{4}}}$

where Gr_L is the Grashof number.

External Flow, Horizontal Plates

W. H. McAdams suggested the following correlations for horizontal plates. The induced buoyancy will be different depending upon whether the hot surface is facing up or down.

For a hot surface facing up or a cold surface facing down:

$$h = \frac{k0.54\text{Ra}_L^{1/4}}{L} \quad 10^5 < \text{Ra}_L < 2\times10^7,$$

$$h = \frac{k0.14\text{Ra}_L^{1/3}}{L} \quad 2\times10^7 < \text{Ra}_L < 3\times10^{10}.$$

For a hot surface facing down or a cold surface facing up:

$$h = \frac{k0.27\text{Ra}_L^{1/4}}{L} \quad 3\times10^5 < \text{Ra}_L < 10^{10}.$$

The characteristic length is the ratio of the plate surface area to perimeter. If the surface is inclined at an angle θ with the vertical then the equations for a vertical plate by Churchill and Chu may be used for θ up to 60°; if the boundary layer flow is laminar, the gravitational constant g is replaced with $g \cos\theta$ when calculating the Ra term.

External Flow, Horizontal Cylinder

For cylinders of sufficient length and negligible end effects, Churchill and Chu has the following correlation for $10^{-5} < \text{Ra}_D < 10^{12}$.

$$h = \frac{k}{D}\left(0.6 + \frac{0.387\text{Ra}_D^{1/6}}{\left(1+(0.559/\text{Pr})^{9/16}\right)^{8/27}}\right)^2$$

External Flow, Spheres

For spheres, T. Yuge has the following correlation for Pr≃1 and $1 \leq \text{Ra}_D \leq 10^5$.

$$\mathrm{Nu}_D = 2 + 0.43\mathrm{Ra}_D^{1/4}$$

Forced Convection

Internal Flow, Laminar Flow

Sieder and Tate has the following correlation for laminar flow in tubes where D is the internal diameter, μ_b is the fluid viscosity at the bulk mean temperature, μ_b is the viscosity at the tube wall surface temperature.

$$Nu_D = 1.86 \cdot (Re \cdot Pr)^{1/3} \left(\frac{D}{L}\right)^{1/3} \left(\frac{\mu_b}{\mu_w}\right)^{0.14}$$

Internal Flow, Turbulent Flow

The Dittus-Bölter correlation (1930) is a common and particularly simple correlation useful for many applications. This correlation is applicable when forced convection is the only mode of heat transfer; i.e., there is no boiling, condensation, significant radiation, etc. The accuracy of this correlation is anticipated to be ±15%.

For a fluid flowing in a straight circular pipe with a Reynolds number between 10,000 and 120,000 (in the turbulent pipe flow range), when the fluid's Prandtl number is between 0.7 and 120, for a location far from the pipe entrance (more than 10 pipe diameters; more than 50 diameters according to many authors) or other flow disturbances, and when the pipe surface is hydraulically smooth, the heat transfer coefficient between the bulk of the fluid and the pipe surface can be expressed explicitly as:

$$\frac{hd}{k} = 0.023\left(\frac{jd}{\mu}\right)^{0.8}\left(\frac{\mu c_p}{k}\right)^n$$

where:

d is the hydraulic diameter

k is the thermal conductivity of the bulk fluid

μ viscosity

j mass flux

c_p isobaric heat capacity

n = 0.4 for heating (wall hotter than the bulk fluid) and 0.33 for cooling (wall cooler than the bulk fluid).

The fluid properties necessary for the application of this equation are evaluated at the bulk temperature thus avoiding iteration

Forced Convection, External Flow

In analyzing the heat transfer associated with the flow past the exterior surface of a solid, the situation is complicated by phenomena such as boundary layer separation. Various authors have correlated charts and graphs for different geometries and flow conditions. For flow parallel to a plane surface, where x is the distance from the edge and L is the height of the boundary layer, a mean Nusselt number can be calculated using the Colburn analogy.

Thom Correlation

There exist simple fluid-specific correlations for heat transfer coefficient in boiling. The Thom correlation is for the flow of boiling water (subcooled or saturated at pressures up to about 20 MPa) under conditions where the nucleate boiling contribution predominates over forced convection. This correlation is useful for rough estimation of expected temperature difference given the heat flux:

$$\Delta T_{sat} = 22.5 \cdot q^{0.5} \exp(-P/8.7)$$

where:

ΔT_{sat} is the wall temperature elevation above the saturation temperature, K

q is the heat flux, MW/m^2

P is the pressure of water, MPa

Note that this empirical correlation is specific to the units given.

Heat Transfer Coefficient of Pipe Wall

The resistance to the flow of heat by the material of pipe wall can be expressed as a "heat transfer coefficient of the pipe wall". However, one needs to select if the heat flux is based on the pipe inner or the outer diameter. Selecting to base the heat flux on the pipe inner diameter, and assuming that the pipe wall thickness is small in comparison with the pipe inner diameter, then the heat transfer coefficient for the pipe wall can be calculated as if the wall were not curved:

$$h_{wall} = \frac{k}{x}$$

where k is the effective thermal conductivity of the wall material and x is the wall thickness.

If the above assumption does not hold, then the wall heat transfer coefficient can be calculated using the following expression:

$$h_{wall} = \frac{2k}{d_i \ln(d_o / d_i)}$$

where d_i and d_o are the inner and outer diameters of the pipe, respectively.

The thermal conductivity of the tube material usually depends on temperature; the mean thermal conductivity is often used.

Combining Heat Transfer Coefficients

For two or more heat transfer processes acting in parallel, heat transfer coefficients simply add:

$$h = h_1 + h_2 + \cdots$$

For two or more heat transfer processes connected in series, heat transfer coefficients add inversely:

$$\frac{1}{h} = \frac{1}{h_1} + \frac{1}{h_2} + \dots$$

For example, consider a pipe with a fluid flowing inside. The approximate rate of heat transfer between the bulk of the fluid inside the pipe and the pipe external surface is:

$$q = \left(\frac{1}{\frac{1}{h} + \frac{t}{k}} \right) \cdot A \cdot \Delta T$$

where

q = heat transfer rate (W)

h = heat transfer coefficient (W/(m^2·K))

t = wall thickness (m)

k = wall thermal conductivity (W/m·K)

A = area (m^2)

ΔT = difference in temperature.

Overall Heat Transfer Coefficient

The overall heat transfer coefficient U is a measure of the overall ability of a series of conductive and convective barriers to transfer heat. It is commonly applied to the calulation of heat transfer in heat exchangers, but can be applied equally well to other problems.

For the case of a heat exchanger, U can be used to determine the total heat transfer between where:

q = heat transfer rate (W)

U = overall heat transfer coefficient (W/(m²·K))

A = heat transfer surface area (m²)

ΔT_{LM} = logarithmic mean temperature difference (K).

The overall heat transfer coefficient takes into account the individual heat transfer coefficients of each stream and the resistance of the pipe material. It can be calculated as the reciprocal of the sum of a series of thermal resistances (but more complex relationships exist, for example when heat transfer takes place by different routes in parallel):

$$\frac{1}{UA} = \sum \frac{1}{hA} + \sum R$$

where:

R = Resistance(s) to heat flow in pipe wall (K/W)

Other parameters are as above.

The heat transfer coefficient is the heat transferred per unit area per kelvin. Thus *area* is included in the equation as it represents the area over which the transfer of heat takes place. The areas for each flow will be different as they represent the contact area for each fluid side.

The *thermal resistance* due to the pipe wall is calculated by the following relationship:

$$R = \frac{x}{k \cdot A}$$

where

x = the wall thickness (m)

k = the thermal conductivity of the material (W/(m·K))

A = the total area of the heat exchanger (m²)

This represents the heat transfer by conduction in the pipe.

The *thermal conductivity* is a characteristic of the particular material. Values of thermal conductivities for various materials are listed in the list of thermal conductivities.

As mentioned earlier in the article the *convection heat transfer coefficient* for each stream depends on the type of fluid, flow properties and temperature properties.

Some typical heat transfer coefficients include:

- Air - h = 10 to 100 W/(m²K)
- Water - h = 500 to 10,000 W/(m²K).

Thermal Resistance Due to Fouling Deposits

Often during their use, heat exchangers collect a layer of fouling on the surface which, in addition to potentially contaminating a stream, reduces the effectiveness of heat exchangers. In a fouled heat exchanger the buildup on the walls creates an additional layer of materials that heat must flow through. Due to this new layer, there is additional resistance within the heat exchanger and thus the overall heat transfer coefficient of the exchanger is reduced. The following relationship is used to solve for the heat transfer resistance with the additional fouling resistance:

$$\frac{1}{U_f P} = \frac{1}{UP} + \frac{R_{fH}}{P_H} + \frac{R_{fC}}{P_C}$$

where

U_f = overall heat transfer coefficient for a fouled heat exchanger, $\frac{W}{m^2 K}$

P = perimeter of the heat exchanger, may be either the hot or cold side perimeter however, it must be the same perimeter on both sides of the equation, m

U = overall heat transfer coefficient for an unfouled heat exchanger, R_{fC}

R_{fC} = fouling resistance on the cold side of the heat exchanger, R_{fC}

R_{fH} = fouling resistance on the hot side of the heat exchanger, $\frac{m^2 K}{W}$

P_C = perimeter of the cold side of the heat exchanger, m

P_H = perimeter of the hot side of the heat exchanger, m

This equation uses the overall heat transfer coefficient of an unfouled heat exchanger and the fouling resistance to calculate the overall heat transfer coefficient of a fouled heat exchanger. The equation takes into account that the perimeter of the heat exchanger is different on the hot and cold sides. The perimeter used for the P does not matter as long as it is the same. The overall heat transfer coefficients will adjust to take into account that a different perimeter was used as the product UP will remain the same.

The fouling resistances can be calculated for a specific heat exchanger if the average thickness and thermal conductivity of the fouling are known. The product of the average thickness and thermal conductivity will result in the fouling resistance on a specific side of the heat exchanger.

$$R_f = \frac{d_f}{k_f}$$

where:

d_f = average thickness of the fouling in a heat exchanger, m

k_f = thermal conductivity of the fouling, $\frac{W}{mK}$..

References

- Paul A., Tipler; Gene Mosca (2008). Physics for Scientists and Engineers, Volume 1 (6th ed.). New York, NY: Worth Publishers. pp. 666–670. ISBN 1-4292-0132-0.
- Raymond Serway; John Jewett (2005), Principles of Physics: A Calculus-Based Text, Cengage Learning, p. 506, ISBN 0-534-49143-X
- Perry, R. H.; Green, D. W., eds. (1997). Perry's Chemical Engineers' Handbook (7th ed.). McGraw-Hill. Table 1–4. ISBN 978-0-07-049841-9.
- Ibach, H.; Luth, H. (2009). Solid-State Physics: An Introduction to Principles of Materials Science. Springer. ISBN 978-3-540-93803-3.
- Frank P. Incropera; David P. De Witt (1990). Fundamentals of Heat and Mass Transfer (3rd ed.). John Wiley & Sons. pp. 100–103. ISBN 0-471-51729-1.
- James R. Welty; Charles E. Wicks; Robert E. Wilson; Gregory L. Rorrer (2007). Fundamentals of Momentum, Heat and Mass transfer (5th edition). John Wiley and Sons. ISBN 978-0470128688.
- Wong, Bill (June 28, 2011), "Drake Landing Solar Community", IDEA/CDEA District Energy/CHP 2011 Conference, Toronto, pp. 1–30, retrieved 21 April 2013

5

Fundamental Theories of Heat Transfer

This chapter aims to stretch the boundaries of physics by clarifying and developing the theoretical and conceptual framework of physics. The fundamental theories such as the first law of thermodynamics, second law of thermodynamics, zeroth law of thermodynamics and NTU method are discussed in this chapter, providing a better understanding of heat transfer.

First Law of Thermodynamics

The first law of thermodynamics is a version of the law of conservation of energy, adapted for thermodynamic systems. The law of conservation of energy states that the total energy of an isolated system is constant; energy can be transformed from one form to another, but cannot be created or destroyed. The first law is often formulated by stating that the change in the internal energy of a closed system is equal to the amount of heat supplied to the system, minus the amount of work done by the system on its surroundings. Equivalently, perpetual motion machines of the first kind are impossible.

History

Investigations into the nature of heat and work and their relationship began with the invention of the first engines used to extract water from mines. Improvements to such engines so as to increase their efficiency and power output came first from mechanics that tinkered with such machines but only slowly advanced the art. Deeper investigations that placed those on a mathematical and physics basis came later.

The process of development of the first law of thermodynamics was by way of much investigative trial and error over a period of about half a century. The first full statements of the law came in 1850 from Rudolf Clausius and from William Rankine; Rankine's statement was perhaps not quite as clear and distinct as was Clausius'. A main aspect of the struggle was to deal with the previously proposed caloric theory of heat.

Germain Hess in 1840 stated a conservation law for the so-called 'heat of reaction' for chemical reactions. His law was later recognized as a consequence of the first law of thermodynamics, but Hess's statement was not explicitly concerned with the relation between energy exchanges by heat and work.

According to Truesdell (1980), Julius Robert von Mayer in 1841 made a statement that meant that "in a process at constant pressure, the heat used to produce expansion is universally interconvertible with work", but this is not a general statement of the first law.

Original Statements: The "Thermodynamic Approach"

The original nineteenth century statements of the first law of thermodynamics appeared in a conceptual framework in which transfer of energy as heat was taken as a primitive notion, not defined or constructed by the theoretical development of the framework, but rather presupposed as prior to it and already accepted. The primitive notion of heat was taken as empirically established, especially through calorimetry regarded as a subject in its own right, prior to thermodynamics. Jointly primitive with this notion of heat were the notions of empirical temperature and thermal equilibrium. This framework also took as primitive the notion of transfer of energy as work. This framework did not presume a concept of energy in general, but regarded it as derived or synthesized from the prior notions of heat and work. By one author, this framework has been called the "thermodynamic" approach.

The first explicit statement of the first law of thermodynamics, by Rudolf Clausius in 1850, referred to cyclic thermodynamic processes.

> *In all cases in which work is produced by the agency of heat, a quantity of heat is consumed which is proportional to the work done; and conversely, by the expenditure of an equal quantity of work an equal quantity of heat is produced.*

Clausius also stated the law in another form, referring to the existence of a function of state of the system, the internal energy, and expressed it in terms of a differential equation for the increments of a thermodynamic process. This equation may described as follows:

> *In a thermodynamic process involving a closed system, the increment in the internal energy is equal to the difference between the heat accumulated by the system and the work done by it.*

Because of its definition in terms of increments, the value of the internal energy of a system is not uniquely defined. It is defined only up to an arbitrary additive constant of integration, which can be adjusted to give arbitrary reference zero levels. This non-uniqueness is in keeping with the abstract mathematical nature of the internal energy. The internal energy is customarily stated relative to a conventionally chosen standard reference state of the system.

The concept of internal energy is considered by Bailyn to be of "enormous interest". Its quantity cannot be immediately measured, but can only be inferred, by differencing actual immediate measurements. Bailyn likens it to the energy states of an atom, that were revealed by Bohr's energy relation $h\nu = E_{n''} - E_{n'}$. In each case, an unmeasurable quantity (the internal energy, the atomic energy level) is revealed by considering the difference of measured quantities (increments of internal energy, quantities of emitted or absorbed radiative energy).

Conceptual Revision: The "Mechanical Approach"

In 1907, George H. Bryan wrote about systems between which there is no transfer of matter (closed systems): "Definition. When energy flows from one system or part of a system to another otherwise than by the performance of mechanical work, the energy so transferred is called *heat*." This definition may be regarded as expressing a conceptual revision, as follows. This was systematically expounded in 1909 by Constantin Carathéodory, whose attention had been drawn to it by Max Born. Largely through Born's influence, this revised conceptual approach to the definition of heat came to be preferred by many twentieth-century writers. It might be called the "mechanical approach".

Energy can also be transferred from one thermodynamic system to another in association with transfer of matter. Born points out that in general such energy transfer is not resolvable uniquely into work and heat moieties. In general, when there is transfer of energy associated with matter transfer, work and heat transfers can be distinguished only when they pass through walls physically separate from those for matter transfer.

The "mechanical" approach postulates the law of conservation of energy. It also postulates that energy can be transferred from one thermodynamic system to another adiabatically as work, and that energy can be held as the internal energy of a thermodynamic system. It also postulates that energy can be transferred from one thermodynamic system to another by a path that is non-adiabatic, and is unaccompanied by matter transfer. Initially, it "cleverly" (according to Bailyn) refrains from labelling as 'heat' such non-adiabatic, unaccompanied transfer of energy. It rests on the primitive notion of *walls*, especially adiabatic walls and non-adiabatic walls, defined as follows. Temporarily, only for purpose of this definition, one can prohibit transfer of energy as work across a wall of interest. Then walls of interest fall into two classes, (a) those such that arbitrary systems separated by them remain independently in their own previously established respective states of internal thermodynamic equilibrium; they are defined as adiabatic; and (b) those without such independence; they are defined as non-adiabatic.

This approach derives the notions of transfer of energy as heat, and of temperature, as theoretical developments, not taking them as primitives. It regards calorimetry as a derived theory. It has an early origin in the nineteenth century, for example in the work of Helmholtz, but also in the work of many others.

Conceptually Revised Statement, According to the Mechanical Approach

The revised statement of the first law postulates that a change in the internal energy of a system due to any arbitrary process, that takes the system from a given initial thermodynamic state to a given final equilibrium thermodynamic state, can be determined through the physical existence, for those given states, of a reference process that occurs purely through stages of adiabatic work.

The revised statement is then

> *For a closed system, in any arbitrary process of interest that takes it from an initial to a final state of internal thermodynamic equilibrium, the change of internal energy is the same as that for a reference adiabatic work process that links those two states. This is so regardless of the path of the process of interest, and regardless of whether it is an adiabatic or a non-adiabatic process. The reference adiabatic work process may be chosen arbitrarily from amongst the class of all such processes.*

This statement is much less close to the empirical basis than are the original statements, but is often regarded as conceptually parsimonious in that it rests only on the concepts of adiabatic work and of non-adiabatic processes, not on the concepts of transfer of energy as heat and of empirical temperature that are presupposed by the original statements. Largely through the influence of Max Born, it is often regarded as theoretically preferable because of this conceptual parsimony. Born particularly observes that the revised approach avoids thinking in terms of what he calls the "imported engineering" concept of heat engines.

Basing his thinking on the mechanical approach, Born in 1921, and again in 1949, proposed to revise the definition of heat. In particular, he referred to the work of Constantin Carathéodory, who had in 1909 stated the first law without defining quantity of heat. Born's definition was specifically for transfers of energy without transfer of matter, and it has been widely followed in textbooks (examples:). Born observes that a transfer of matter between two systems is accompanied by a transfer of internal energy that cannot be resolved into heat and work components. There can be pathways to other systems, spatially separate from that of the matter transfer, that allow heat and work transfer independent of and simultaneous with the matter transfer. Energy is conserved in such transfers.

Description

The first law of thermodynamics for a closed system was expressed in two ways by Clausius. One way referred to cyclic processes and the inputs and outputs of the system, but did not refer to increments in the internal state of the system. The other way referred to an incremental change in the internal state of the system, and did not expect the process to be cyclic.

A cyclic process is one that can be repeated indefinitely often, returning the system to its initial state. Of particular interest for single cycle of a cyclic process are the net work done, and the net heat taken in (or 'consumed', in Clausius' statement), by the system.

In a cyclic process in which the system does net work on its surroundings, it is observed to be physically necessary not only that heat be taken into the system, but also, importantly, that some heat leave the system. The difference is the heat converted by the cycle into work. In each repetition of a cyclic process, the net work done by the system, measured in mechanical units, is proportional to the heat consumed, measured in calorimetric units.

The constant of proportionality is universal and independent of the system and in 1845 and 1847 was measured by James Joule, who described it as the *mechanical equivalent of heat.*

In a non-cyclic process, the change in the internal energy of a system is equal to net energy added as heat to the system minus the net work done by the system, both being measured in mechanical units. Taking ΔU as a change in internal energy, one writes

$$\Delta U = Q - W \text{ (sign convention of Clausius and generallyin this article)}$$

where Q denotes the net quantity of heat supplied to the system by its surroundings and W denotes the net work done by the system. This sign convention is implicit in Clausius' statement of the law given above. It originated with the study of heat engines that produce useful work by consumption of heat.

Often nowadays, however, writers use the IUPAC convention by which the first law is formulated with work done on the system by its surroundings having a positive sign. With this now often used sign convention for work, the first law for a closed system may be written:

$$\Delta U = Q + W \text{ (sign convention of IUPAC).}$$

This convention follows physicists such as Max Planck, and considers all net energy transfers to the system as positive and all net energy transfers from the system as negative, irrespective of any use for the system as an engine or other device.

When a system expands in a fictive quasistatic process, the work done by the system on the environment is the product, $P\,\mathrm{d}V$, of pressure, P, and volume change, $\mathrm{d}V$, whereas the work done *on* the system is $-P\,\mathrm{d}V$. Using either sign convention for work, the change in internal energy of the system is:

$$\mathrm{d}U = \delta Q - P\mathrm{d}V \text{ (quasi-static process),}$$

where δQ denotes the infinitesimal increment of heat supplied to the system from its surroundings.

Work and heat are expressions of actual physical processes of supply or removal of energy, while the internal energy U is a mathematical abstraction that keeps account of the exchanges of energy that befall the system. Thus the term heat for Q means "that amount of energy added or removed by conduction of heat or by thermal radiation", rather than referring to a form of energy within the system. Likewise, the term work energy for W means "that amount of energy gained or lost as the result of work". Internal energy is a property of the system whereas work done and heat supplied are not. A significant result of this distinction is that a given internal energy change ΔU can be achieved by, in principle, many combinations of heat and work.

Various Statements of the Law for Closed Systems

The law is of great importance and generality and is consequently thought of from several points of view. Most careful textbook statements of the law express it for closed systems. It is stated in several ways, sometimes even by the same author.

For the thermodynamics of closed systems, the distinction between transfers of energy as work and as heat is central and is within the scope of the present article. For the thermodynamics of open systems, such a distinction is beyond the scope of the present article, but some limited comments are made on it in the section below headed 'First law of thermodynamics for open systems'.

There are two main ways of stating a law of thermodynamics, physically or mathematically. They should be logically coherent and consistent with one another.

An example of a physical statement is that of Planck (1897/1903):

> It is in no way possible, either by mechanical, thermal, chemical, or other devices, to obtain perpetual motion, i.e. it is impossible to construct an engine which will work in a cycle and produce continuous work, or kinetic energy, from nothing.

This physical statement is restricted neither to closed systems nor to systems with states that are strictly defined only for thermodynamic equilibrium; it has meaning also for open systems and for systems with states that are not in thermodynamic equilibrium.

An example of a mathematical statement is that of Crawford (1963):

> For a given system we let ΔE^{kin} = large-scale mechanical energy, ΔE^{pot} = large-scale potential energy, and ΔE^{tot} = total energy. The first two quantities are specifiable in terms of appropriate mechanical variables, and by definition
>
> $$E^{\text{tot}} = E^{\text{kin}} + E^{\text{pot}} + U.$$
>
> For any finite process, whether reversible or irreversible,
>
> $$\Delta E^{\text{tot}} = \Delta E^{\text{kin}} + \Delta E^{\text{pot}} + \Delta U.$$
>
> The first law in a form that involves the principle of conservation of energy more generally is
>
> $$\Delta E^{\text{tot}} = Q + W.$$
>
> Here Q and W are heat and work added, with no restrictions as to whether the process is reversible, quasistatic, or irreversible.[Warner, *Am. J. Phys.*, 29, 124 (1961)]

This statement by Crawford, for W, uses the sign convention of IUPAC, not that of Clausius. Though it does not explicitly say so, this statement refers to closed systems, and to internal energy U defined for bodies in states of thermodynamic equilibrium, which possess well-defined temperatures.

The history of statements of the law for closed systems has two main periods, before and after the work of Bryan (1907), of Carathéodory (1909), and the approval of Carathéodory's work given by Born (1921). The earlier traditional versions of the law for closed systems are nowadays often considered to be out of date.

Carathéodory's celebrated presentation of equilibrium thermodynamics refers to closed systems, which are allowed to contain several phases connected by internal walls of various kinds of impermeability and permeability (explicitly including walls that are permeable only to heat). Carathéodory's 1909 version of the first law of thermodynamics was stated in an axiom which refrained from defining or mentioning temperature or quantity of heat transferred. That axiom stated that the internal energy of a phase in equilibrium is a function of state, that the sum of the internal energies of the phases is the total internal energy of the system, and that the value of the total internal energy of the system is changed by the amount of work done adiabatically on it, considering work as a form of energy. That article considered this statement to be an expression of the law of conservation of energy for such systems. This version is nowadays widely accepted as authoritative, but is stated in slightly varied ways by different authors.

Such statements of the first law for closed systems assert the existence of internal energy as a function of state defined in terms of adiabatic work. Thus heat is not defined calorimetrically or as due to temperature difference. It is defined as a residual difference between change of internal energy and work done on the system, when that work does not account for the whole of the change of internal energy and the system is not adiabatically isolated.

The 1909 Carathéodory statement of the law in axiomatic form does not mention heat or temperature, but the equilibrium states to which it refers are explicitly defined by variable sets that necessarily include "non-deformation variables", such as pressures, which, within reasonable restrictions, can be rightly interpreted as empirical temperatures, and the walls connecting the phases of the system are explicitly defined as possibly impermeable to heat or permeable only to heat.

According to Münster (1970), "A somewhat unsatisfactory aspect of Carathéodory's theory is that a consequence of the Second Law must be considered at this point [in the statement of the first law], i.e. that it is not always possible to reach any state 2 from any other state 1 by means of an adiabatic process." Münster instances that no adiabatic process can reduce the internal energy of a system at constant volume. Carathéodory's paper asserts that its statement of the first law corresponds exactly to Joule's experimental arrangement, regarded as an instance of adiabatic work. It does not point out that Joule's experimental arrangement performed essentially irreversible work, through friction of paddles in a liquid, or passage of electric current through a resistance inside the system, driven by motion of a coil and inductive heating, or by an external current source, which can access the system only by the passage of electrons, and so is not strictly adiabatic, because electrons are a form of matter, which cannot penetrate adiabatic walls. The paper goes on to base its main argument on the possibility of quasi-static adiabatic work, which is essentially reversible. The paper asserts that it will avoid reference to Carnot cycles, and then proceeds to base its argument on cycles of forward and backward quasi-static adiabatic stages, with isothermal stages of zero magnitude.

Sometimes the concept of internal energy is not made explicit in the statement.

Sometimes the existence of the internal energy is made explicit but work is not explicitly mentioned in the statement of the first postulate of thermodynamics. Heat supplied is then defined as the residual change in internal energy after work has been taken into account, in a non-adiabatic process.

A respected modern author states the first law of thermodynamics as "Heat is a form of energy", which explicitly mentions neither internal energy nor adiabatic work. Heat is defined as energy transferred by thermal contact with a reservoir, which has a temperature, and is generally so large that addition and removal of heat do not alter its temperature. A current student text on chemistry defines heat thus: "*heat* is the exchange of thermal energy between a system and its surroundings caused by a temperature difference." The author then explains how heat is defined or measured by calorimetry, in terms of heat capacity, specific heat capacity, molar heat capacity, and temperature.

A respected text disregards the Carathéodory's exclusion of mention of heat from the statement of the first law for closed systems, and admits heat calorimetrically defined along with work and internal energy. Another respected text defines heat exchange as determined by temperature difference, but also mentions that the Born (1921) version is "completely rigorous". These versions follow the traditional approach that is now considered out of date, exemplified by that of Planck (1897/1903).

Evidence for the First Law of Thermodynamics for Closed Systems

The first law of thermodynamics for closed systems was originally induced from empirically observed evidence, including calorimetric evidence. It is nowadays, however, taken to provide the definition of heat via the law of conservation of energy and the definition of work in terms of changes in the external parameters of a system. The original discovery of the law was gradual over a period of perhaps half a century or more, and some early studies were in terms of cyclic processes.

The following is an account in terms of changes of state of a closed system through compound processes that are not necessarily cyclic. This account first considers processes for which the first law is easily verified because of their simplicity, namely adiabatic processes (in which there is no transfer as heat) and adynamic processes (in which there is no transfer as work).

Adiabatic Processes

In an adiabatic process, there is transfer of energy as work but not as heat. For all adiabatic process that takes a system from a given initial state to a given final state, irrespective of how the work is done, the respective eventual total quantities of energy transferred as work are one and the same, determined just by the given initial and final states. The work done on the system is defined and measured by changes in mechanical or quasi-mechanical variables external to the system. Physically, adiabatic transfer of energy as work requires the existence of adiabatic enclosures.

For instance, in Joule's experiment, the initial system is a tank of water with a paddle wheel inside. If we isolate the tank thermally, and move the paddle wheel with a pulley and a weight, we can relate the increase in temperature with the distance descended by the mass. Next, the system is returned to its initial state, isolated again, and the same amount of work is done on the tank using different devices (an electric motor, a chemical battery, a spring,...). In every case, the amount of work can be measured independently. The return to the initial state is not conducted by doing adiabatic work on the system. The evidence shows that the final state of the water (in particular, its temperature and volume) is the same in every case. It is irrelevant if the work is electrical, mechanical, chemical,... or if done suddenly or slowly, as long as it is performed in an adiabatic way, that is to say, without heat transfer into or out of the system.

Evidence of this kind shows that to increase the temperature of the water in the tank, the qualitative kind of adiabatically performed work does not matter. No qualitative kind of adiabatic work has ever been observed to decrease the temperature of the water in the tank.

A change from one state to another, for example an increase of both temperature and volume, may be conducted in several stages, for example by externally supplied electrical work on a resistor in the body, and adiabatic expansion allowing the body to do work on the surroundings. It needs to be shown that the time order of the stages, and their relative magnitudes, does not affect the amount of adiabatic work that needs to be done for the change of state. According to one respected scholar: "Unfortunately, it does not seem that experiments of this kind have ever been carried out carefully. ... We must therefore admit that the statement which we have enunciated here, and which is equivalent to the first law of thermodynamics, is not well founded on direct experimental evidence." Another expression of this view is "... no systematic precise experiments to verify this generalization directly have ever been attempted."

This kind of evidence, of independence of sequence of stages, combined with the above-mentioned evidence, of independence of qualitative kind of work, would show the existence of an important state variable that corresponds with adiabatic work, but not that such a state variable represented a conserved quantity. For the latter, another step of evidence is needed, which may be related to the concept of reversibility, as mentioned below.

That important state variable was first recognized and denoted U by Clausius in 1850, but he did not then name it, and he defined it in terms not only of work but also of heat transfer in the same process. It was also independently recognized in 1850 by Rankine, who also denoted it U ; and in 1851 by Kelvin who then called it "mechanical energy", and later "intrinsic energy". In 1865, after some hestitation, Clausius began calling his state function U "energy". In 1882 it was named as the *internal energy* by Helmholtz. If only adiabatic processes were of interest, and heat could be ignored, the concept of internal energy would hardly arise or be needed. The relevant physics would be largely covered by the concept of potential energy, as was intended in the 1847 paper of Helmholtz on the principle of conservation of energy, though that did not deal with forces that cannot be described by a potential, and thus did not fully justify the principle. Moreover, that paper was critical of the early work of Joule that had by then been performed. A great merit of the internal energy concept is that it frees thermodynamics from a restriction to cyclic processes, and allows a treatment in terms of thermodynamic states.

In an adiabatic process, adiabatic work takes the system either from a reference state O with internal energy $U(O)$ to an arbitrary one A with internal energy $U(A)$, or from the state A to the state O:

$$U(A) = U(O) - W_{O\to A}^{\text{adiabatic}} \text{ or } U(O) = U(A) - W_{A\to O}^{\text{adiabatic}}.$$

Except under the special, and strictly speaking, fictional, condition of reversibility, only one of the processes adiabatic, $O \to A$ or adiabatic, $A \to O$ is empirically feasible by a simple application of externally supplied work. The reason for this is given as the second law of thermodynamics and is not considered in the present article.

The fact of such irreversibility may be dealt with in two main ways, according to different points of view:

- Since the work of Bryan (1907), the most accepted way to deal with it nowadays, followed by Carathéodory, is to rely on the previously established concept of quasi-static processes, as follows. Actual physical processes of transfer of energy as work are always at least to some degree irreversible. The irreversibility is often due to mechanisms known as dissipative, that transform bulk kinetic energy into internal energy. Examples are friction and viscosity. If the process is performed more slowly, the frictional or viscous dissipation is less. In the limit of infinitely slow performance, the dissipation tends to zero and then the limiting process, though fictional rather than actual, is notionally reversible, and is called quasi-static. Throughout the course of the fictional limiting quasi-static process, the internal intensive variables of the system are equal to the external intensive variables, those that describe the reactive forces exerted by the surroundings. This can be taken to justify the formula

$$(1)\ W_{A\to O}^{\text{adiabatic, quasi-static}} = -W_{O\to A}^{\text{adiabatic, quasi-static}}.$$

- Another way to deal with it is to allow that experiments with processes of heat transfer to or from the system may be used to justify the formula (1) above. Moreover, it deals to some extent with the problem of lack of direct experimental evidence that the time order of stages of a process does not matter in the determination of internal energy. This way does not provide theoretical purity in terms of adiabatic work processes, but is empirically feasible, and is in accord with experiments actually done, such as the Joule experiments mentioned just above, and with older traditions.

The formula (1) above allows that to go by processes of quasi-static adiabatic work from the state A to the state B we can take a path that goes through the reference state O, since the quasi-static adiabatic work is independent of the path

$$-W_{A\to B}^{\text{adiabatic,quasi-static}} = -W_{A\to O}^{\text{adiabatic,quasi-static}} - W_{O\to B}^{\text{adiabatic,quasi-static}} = W_{O\to A}^{\text{adiabatic,quasi-static}} - W_{O\to B}^{\text{adiabatic,quasi-static}}$$

$$= -U(A) + U(B) = \Delta U$$

This kind of empirical evidence, coupled with theory of this kind, largely justifies the following statement:

> *For all adiabatic processes between two specified states of a closed system of any nature, the net work done is the same regardless the details of the process, and determines a state function called internal energy, U.*"

Adynamic Processes

A complementary observable aspect of the first law is about heat transfer. Adynamic transfer of energy as heat can be measured empirically by changes in the surroundings of the system of interest by calorimetry. This again requires the existence of adiabatic enclosure of the entire process, system and surroundings, though the separating wall between the surroundings and the system is thermally conductive or radiatively permeable, not adiabatic. A calorimeter can rely on measurement of sensible heat, which requires the existence of thermometers and measurement of temperature change in bodies of known sensible heat capacity under specified conditions; or it can rely on the measurement of latent heat, through measurement of masses of material that change phase, at temperatures fixed by the occurrence of phase changes under specified conditions in bodies of known latent heat of phase change. The calorimeter can be calibrated by adiabatically doing externally determined work on it. The most accurate method is by passing an electric current from outside through a resistance inside the calorimeter. The calibration allows comparison of calorimetric measurement of quantity of heat transferred with quantity of energy transferred as work. According to one textbook, "The most common device for measuring ΔU is an adiabatic bomb calorimeter." According to another textbook, "Calorimetry is widely used in present day laboratories." According to one opinion, "Most thermodynamic data come from calorimetry..." According to another opinion, "The most common method of measuring "heat" is with a calorimeter."

When the system evolves with transfer of energy as heat, without energy being transferred as work, in an adynamic process, the heat transferred to the system is equal to the increase in its internal energy:

$$Q_{A\to B}^{\text{adynamic}} = \Delta U.$$

General Case for Reversible Processes

Heat transfer is practically reversible when it is driven by practically negligibly small temperature gradients. Work transfer is practically reversible when it occurs so slowly that there are no frictional effects within the system; frictional effects outside the system should also be zero if the process is to be globally reversible. For a particular reversible process in general, the work done reversibly on the system, $W_{A\to B}^{\text{path } P_0,\text{reversible}}$ and the heat transferred reversibly to the system , $Q_{A\to B}^{\text{path } P_0,\text{reversible}}$ are not required to occur respectively adiabatically or adynamically, but they must belong to the same particular process defined by its particular reversible path, , through the space of thermodynamic states. Then the work and heat transfers can occur and be calculated simultaneously.

Putting the two complementary aspects together, the first law for a particular reversible process can be written

$$-W_{A\to B}^{\text{path } P_0,\text{reversible}} + Q_{A\to B}^{\text{path } P_0,\text{reversible}} = \Delta U.$$

This combined statement is the expression the first law of thermodynamics for reversible processes for closed systems.

In particular, if no work is done on a thermally isolated closed system we have

$$\Delta U = 0.$$

This is one aspect of the law of conservation of energy and can be stated:

The internal energy of an isolated system remains constant.

General Case for Irreversible Processes

If, in a process of change of state of a closed system, the energy transfer is not under a practically zero temperature gradient and practically frictionless, then the process is irreversible. Then the heat and work transfers may be difficult to calculate, and irreversible thermodynamics is called for. Nevertheless, the first law still holds and provides a check on the measurements and calculations of the work done irreversibly on the system, $W_{A \to B}^{\text{path } P_1, \text{irreversible}}$, and the heat transferred irreversibly to the system, $Q_{A \to B}^{\text{path } P_1, \text{irreversible}}$, which belong to the same particular process defined by its particular irreversible path, P_1, through the space of thermodynamic states.

$$-W_{A \to B}^{\text{path } P_1, \text{irreversible}} + Q_{A \to B}^{\text{path } P_1, \text{irreversible}} = \Delta U.$$

This means that the internal energy U is a function of state and that the internal energy change ÄU between two states is a function only of the two states.

Overview of the Weight of Evidence for the Law

The first law of thermodynamics is so general that its predictions cannot all be directly tested. In many properly conducted experiments it has been precisely supported, and never violated. Indeed, within its scope of applicability, the law is so reliably established, that, nowadays, rather than experiment being considered as testing the accuracy of the law, it is more practical and realistic to think of the law as testing the accuracy of experiment. An experimental result that seems to violate the law may be assumed to be inaccurate or wrongly conceived, for example due to failure to account for an important physical factor. Thus, some may regard it as a principle more abstract than a law.

State Functional Formulation for Infinitesimal Processes

When the heat and work transfers in the equations above are infinitesimal in magnitude, they are often denoted by δ, rather than exact differentials denoted by d, as a reminder that heat and work do not describe the *state* of any system. The integral of an inexact differential depends upon the particular path taken through the space of thermodynamic parameters while the integral of an exact differential depends only upon the initial and final states. If the initial and final states are the same, then the integral of an inexact differential may or may not be zero, but the integral of an exact differential is always zero. The path taken by a thermodynamic system through a chemical or physical change is known as a thermodynamic process.

The first law for a closed homogeneous system may be stated in terms that include concepts that are established in the second law. The internal energy U may then be expressed as a function of the system's defining state variables S, entropy, and V, volume: $U = U(S, V)$. In these terms, T, the system's temperature, and P, its pressure, are partial derivatives of U with respect to S and V. These variables are important throughout thermodynamics, though not necessary for the statement of the first law. Rigorously, they are defined only when the system is in its own state of internal thermodynamic equilibrium. For some purposes, the concepts provide good approximations for scenarios sufficiently near to the system's internal thermodynamic equilibrium.

The first law requires that:

$$dU = \delta Q - \delta W \qquad \text{(closed system, general process, quasi-static or irreversible)}$$

Then, for the fictive case of a reversible process, dU can be written in terms of exact differentials. One may imagine reversible changes, such that there is at each instant negligible departure from thermodynamic equilibrium within the system. This excludes isochoric work. Then, mechanical work is given by $\delta W = -P\,dV$ and the quantity of heat added can be expressed as $\delta Q = T\,dS$. For these conditions

$$dU = Tds - PdV \qquad \text{(closed system, reversible process).}$$

While this has been shown here for reversible changes, it is valid in general, as U can be considered as a thermodynamic state function of the defining state variables S and V:

$$\text{(2)}\quad dU = TdS - PdV \quad \text{(closed system, general process, quasi-static or irreversible).}$$

Equation (2) is known as the fundamental thermodynamic relation for a closed system in the energy representation, for which the defining state variables are S and V, with respect to which T and P are partial derivatives of U. It is only in the fictive reversible case, when isochoric work is excluded, that the work done and heat transferred are given by $-P\,\mathrm{d}V$ and $T\,\mathrm{d}S$.

In the case of a closed system in which the particles of the system are of different types and, because chemical reactions may occur, their respective numbers are not necessarily constant, the fundamental thermodynamic relation for dU becomes:

$$dU = TdS - PdV + \sum_i \mu_i dN_i.$$

where $\mathrm{d}N_i$ is the (small) increase in amount of type-i particles in the reaction, and μ_i is known as the chemical potential of the type-i particles in the system. If $\mathrm{d}N_i$ is expressed in mol then μ_i is expressed in J/mol. If the system has more external mechanical variables than just the volume that can change, the fundamental thermodynamic relation further generalizes to:

$$dU = TdS - \sum_i X_i dx_i + \sum_j \mu_j dN_j.$$

Here the X_i are the generalized forces corresponding to the external variables x_i. The parameters X_i are independent of the size of the system and are called intensive parameters and the x_i are proportional to the size and called extensive parameters.

For an open system, there can be transfers of particles as well as energy into or out of the system during a process. For this case, the first law of thermodynamics still holds, in the form that the internal energy is a function of state and the change of internal energy in a process is a function only of its initial and final states, as noted in the section below headed First law of thermodynamics for open systems.

A useful idea from mechanics is that the energy gained by a particle is equal to the force applied to the particle multiplied by the displacement of the particle while that force is applied. Now consider the first law without the heating term: $dU = -PdV$. The pressure P can be viewed as a force (and in fact has units of force per unit area) while dV is the displacement (with units of distance times area). We may say, with respect to this work term, that a pressure difference forces a transfer of volume, and that the product of the two (work) is the amount of energy transferred out of the system as a result of the process. If one were to make this term negative then this would be the work done on the system.

It is useful to view the TdS term in the same light: here the temperature is known as a "generalized" force (rather than an actual mechanical force) and the entropy is a generalized displacement.

Similarly, a difference in chemical potential between groups of particles in the system drives a chemical reaction that changes the numbers of particles, and the corresponding product is the amount of chemical potential energy transformed in process. For example, consider a system consisting of two phases: liquid water and water vapor. There is a generalized "force" of evaporation that drives water molecules out of the liquid. There is a generalized "force" of condensation that drives vapor molecules out of the vapor. Only when these two "forces" (or chemical potentials) are equal is there equilibrium, and the net rate of transfer zero.

The two thermodynamic parameters that form a generalized force-displacement pair are called "conjugate variables". The two most familiar pairs are, of course, pressure-volume, and temperature-entropy.

Spatially Inhomogeneous Systems

Classical thermodynamics is initially focused on closed homogeneous systems (e.g. Planck 1897/1903), which might be regarded as 'zero-dimensional' in the sense that they have no spatial variation. But it is desired to study also systems with distinct internal motion and spatial inhomogeneity. For such systems, the principle of conservation of energy is expressed in terms not only of internal energy as defined for homogeneous systems, but also in terms of kinetic energy and potential energies of parts of the inhomogeneous system with respect to each other and with respect to long-range external forces. How the total energy of a system is allocated between these three more specific kinds of energy varies according to the purposes of different writers; this is because these components of energy are to some extent mathematical artefacts rather than actually measured physical quantities. For any closed homogeneous component of an inhomogeneous closed system, if E denotes the total energy of that component system, one may write

$$E = E^{\mathrm{kin}} + E^{\mathrm{pot}} + U$$

where E^{kin} and E^{pot} denote respectively the total kinetic energy and the total potential energy of the component closed homogeneous system, and U denotes its internal energy.

Potential energy can be exchanged with the surroundings of the system when the surroundings impose a force field, such as gravitational or electromagnetic, on the system.

A compound system consisting of two interacting closed homogeneous component subsystems has a potential energy of interaction E_{12}^{pot} between the subsystems. Thus, in an obvious notation, one may write

$$E = E_1^{\text{kin}} + E_1^{\text{pot}} + U_1 + E_2^{\text{kin}} + E_2^{\text{pot}} + U_2 + E_{12}^{\text{pot}}$$

The quantity E_{12}^{pot} in general lacks an assignment to either subsystem in a way that is not arbitrary, and this stands in the way of a general non-arbitrary definition of transfer of energy as work. On occasions, authors make their various respective arbitrary assignments.

The distinction between internal and kinetic energy is hard to make in the presence of turbulent motion within the system, as friction gradually dissipates macroscopic kinetic energy of localised bulk flow into molecular random motion of molecules that is classified as internal energy. The rate of dissipation by friction of kinetic energy of localised bulk flow into internal energy, whether in turbulent or in streamlined flow, is an important quantity in non-equilibrium thermodynamics. This is a serious difficulty for attempts to define entropy for time-varying spatially inhomogeneous systems.

First Law of Thermodynamics for Open Systems

For the first law of thermodynamics, there is no trivial passage of physical conception from the closed system view to an open system view. For closed systems, the concepts of an adiabatic enclosure and of an adiabatic wall are fundamental. Matter and internal energy cannot permeate or penetrate such a wall. For an open system, there is a wall that allows penetration by matter. In general, matter in diffusive motion carries with it some internal energy, and some microscopic potential energy changes accompany the motion. An open system is not adiabatically enclosed.

There are some cases in which a process for an open system can, for particular purposes, be considered as if it were for a closed system. In an open system, by definition hypothetically or potentially, matter can pass between the system and its surroundings. But when, in a particular case, the process of interest involves only hypothetical or potential but no actual passage of matter, the process can be considered as if it were for a closed system.

Internal Energy for an Open System

Since the revised and more rigorous definition of the internal energy of a closed system rests upon the possibility of processes by which adiabatic work takes the system from one state to another, this leaves a problem for the definition of internal energy for an open system, for which adiabatic work is not in general possible. According to Max Born, the transfer of matter and energy across an open connection "cannot be reduced to mechanics". In contrast to the case of closed systems, for open systems, in the presence of diffusion, there is no unconstrained and unconditional physical distinction between convective transfer of internal energy by bulk flow of matter, the transfer of internal energy without transfer of matter (usually called heat conduction and work transfer), and change of various potential energies. The older traditional way and the conceptually revised (Carathéodory) way agree that there is no physically unique definition of heat and work transfer processes between open systems.

In particular, between two otherwise isolated open systems an adiabatic wall is by definition impossible. This problem is solved by recourse to the principle of conservation of energy. This principle allows a composite isolated system to be derived from two other component non-interacting isolated systems, in such a way that the total energy of the composite isolated system is equal to the sum of the total energies of the two component isolated systems. Two previously isolated systems can be subjected to the thermodynamic operation of placement between them of a wall permeable to matter and energy, followed by a time for establishment of a new thermodynamic state of internal equilibrium in the new single unpartitioned system. The internal energies of the initial two systems and of the final new system, considered respectively as closed systems as above, can be measured. Then the law of conservation of energy requires that

$$\Delta U_s + \Delta U_o = 0,$$

where ΔU_s and ΔU_o denote the changes in internal energy of the system and of its surroundings respectively. This is a statement of the first law of thermodynamics for a transfer between two otherwise isolated open systems, that fits well with the conceptually revised and rigorous statement of the law stated above.

For the thermodynamic operation of adding two systems with internal energies U_1 and U_2, to produce a new system with internal energy U, one may write $U = U_1 + U_2$; the reference states for U, U_1 and U_2 should be specified accordingly, maintaining also that the internal energy of a system be proportional to its mass, so that the internal energies are extensive variables.

There is a sense in which this kind of additivity expresses a fundamental postulate that goes beyond the simplest ideas of classical closed system thermodynamics; the extensivity of some variables is not obvious, and needs explicit expression; indeed one author goes so far as to say that it could be recognized as a fourth law of thermodynamics, though this is not repeated by other authors.

Also of course

$$\Delta N_s + \Delta N_o = 0,$$

where ΔN_s and ΔN_o denote the changes in mole number of a component substance of the system and of its surroundings respectively. This is a statement of the law of conservation of mass.

Process of Transfer of Matter between an Open System and its Surroundings

A system connected to its surroundings only through contact by a single permeable wall, but otherwise isolated, is an open system. If it is initially in a state of contact equilibrium with a surrounding subsystem, a thermodynamic process of transfer of matter can be made to occur between them if the surrounding subsystem is subjected to some thermodynamic operation, for example, removal of a partition between it and some further surrounding subsystem. The removal of the partition in the surroundings initiates a process of exchange between the system and its contiguous surrounding subsystem.

An example is evaporation. One may consider an open system consisting of a collection of liquid, enclosed except where it is allowed to evaporate into or to receive condensate from its vapor above it, which may be considered as its contiguous surrounding subsystem, and subject to control of its volume and temperature.

A thermodynamic process might be initiated by a thermodynamic operation in the surroundings, that mechanically increases in the controlled volume of the vapor. Some mechanical work will be done within the surroundings by the vapor, but also some of the parent liquid will evaporate and enter the vapor collection which is the contiguous surrounding subsystem. Some internal energy will accompany the vapor that leaves the system, but it will not make sense to try to uniquely identify part of that internal energy as heat and part of it as work. Consequently, the energy transfer that accompanies the transfer of matter between the system and its surrounding subsystem cannot be uniquely split into heat and work transfers to or from the open system. The component of total energy transfer that accompanies the transfer of vapor into the surrounding subsystem is customarily called 'latent heat of evaporation', but this use of the word heat is a quirk of customary historical language, not in strict compliance with the thermodynamic definition of transfer of energy as heat. In this example, kinetic energy of bulk flow and potential energy with respect to long-range external forces such as gravity are both considered to be zero. The first law of thermodynamics refers to the change of internal energy of the open system, between its initial and final states of internal equilibrium.

Open System with Multiple Contacts

An open system can be in contact equilibrium with several other systems at once.

This includes cases in which there is contact equilibrium between the system, and several subsystems in its surroundings, including separate connections with subsystems through walls that are permeable to the transfer of matter and internal energy as heat and allowing friction of passage of the transferred matter, but immovable, and separate connections through adiabatic walls with others, and separate connections through diathermic walls impermeable to matter with yet others. Because there are physically separate connections that are permeable to energy but impermeable to matter, between the system and its surroundings, energy transfers between them can occur with definite heat and work characters. Conceptually essential here is that the internal energy transferred with the transfer of matter is measured by a variable that is mathematically independent of the variables that measure heat and work.

With such independence of variables, the total increase of internal energy in the process is then determined as the sum of the internal energy transferred from the surroundings with the transfer of matter through the walls that are permeable to it, and of the internal energy transferred to the system as heat through the diathermic walls, and of the energy transferred to the system as work through the adiabatic walls, including the energy transferred to the system by long-range forces. These simultaneously transferred quantities of energy are defined by events in the surroundings of the system. Because the internal energy transferred with matter is not in general uniquely resolvable into heat and work components, the total energy transfer cannot in general be uniquely resolved into heat and work components. Under these conditions, the following formula can describe the process in terms of externally defined thermodynamic variables, as a statement of the first law of thermodynamics:

$$(3)\quad \Delta U_0 = Q - W - \sum_{i=1}^{m} \Delta U_i$$

(suitably defined surrounding subsystems, general process, quasi-static or irreversible),

where ΔU_0 denotes the change of internal energy of the system, and ΔU_i denotes the change of internal energy of the *i*th of the *m* surrounding subsystems that are in open contact with the system, due to transfer between the system and that *i*th surrounding subsystem, and Q denotes the internal energy transferred as heat from the heat reservoir of the surroundings to the system, and W denotes the energy transferred from the system to the surrounding subsystems that are in adiabatic connection with it. The case of a wall that is permeable to matter and can move so as to allow transfer of energy as work is not considered here.

Combination of First and Second Laws

If the system is described by the energetic fundamental equation, $U_0 = U_0(S, V, N_j)$, and if the process can be described in the quasi-static formalism, in terms of the internal state variables of the system, then the process can also be described by a combination of the first and second laws of thermodynamics, by the formula

$$(4)\ \mathrm{d}U_0 = T\mathrm{d}S - P\mathrm{d}V + \sum_{j=1}^{n} \mu_j \mathrm{d}N_j$$

where there are n chemical constituents of the system and permeably connected surrounding subsystems, and where T, S, P, V, N_j, and μ_j, are defined as above.

For a general natural process, there is no immediate term-wise correspondence between equations (3) and (4), because they describe the process in different conceptual frames.

Nevertheless, a conditional correspondence exists. There are three relevant kinds of wall here: purely diathermal, adiabatic, and permeable to matter. If two of those kinds of wall are sealed off, leaving only one that permits transfers of energy, as work, as heat, or with matter, then the remaining permitted terms correspond precisely. If two of the kinds of wall are left unsealed, then energy transfer can be shared between them, so that the two remaining permitted terms do not correspond precisely.

For the special fictive case of quasi-static transfers, there is a simple correspondence. For this, it is supposed that the system has multiple areas of contact with its surroundings. There are pistons that allow adiabatic work, purely diathermal walls, and open connections with surrounding subsystems of completely controllable chemical potential (or equivalent controls for charged species). Then, for a suitable fictive quasi-static transfer, one can write

$$\delta Q = T\mathrm{d}S \text{ and } \delta W = P\mathrm{d}V$$

(suitably defined surrounding subsystems, quasi-static transfers of energy).

For fictive quasi-static transfers for which the chemical potentials in the connected surrounding subsystems are suitably controlled, these can be put into equation (4) to yield

$$(5)\ \mathrm{d}U_0 = \delta Q - \delta W + \sum_{j=1}^{n} \mu_j \mathrm{d}N_j$$

(suitably defined surrounding subsystems, quasi-static transfers).

The reference does not actually write equation (5), but what it does write is fully compatible with it. Another helpful account is given by Tschoegl.

There are several other accounts of this, in apparent mutual conflict.

Non-equilibrium Transfers

The transfer of energy between an open system and a single contiguous subsystem of its surroundings is considered also in non-equilibrium thermodynamics. The problem of definition arises also in this case. It may be allowed that the wall between the system and the subsystem is not only permeable to matter and to internal energy, but also may be movable so as to allow work to be done when the two systems have different pressures. In this case, the transfer of energy as heat is not defined.

Methods for study of non-equilibrium processes mostly deal with spatially continuous flow systems. In this case, the open connection between system and surroundings is usually taken to fully surround the system, so that there are no separate connections impermeable to matter but permeable to heat. Except for the special case mentioned above when there is no actual transfer of matter, which can be treated as if for a closed system, in strictly defined thermodynamic terms, it follows that transfer of energy as heat is not defined. In this sense, there is no such thing as 'heat flow' for a continuous-flow open system. Properly, for closed systems, one speaks of transfer of internal energy as heat, but in general, for open systems, one can speak safely only of transfer of internal energy. A factor here is that there are often cross-effects between distinct transfers, for example that transfer of one substance may cause transfer of another even when the latter has zero chemical potential gradient.

Usually transfer between a system and its surroundings applies to transfer of a state variable, and obeys a balance law, that the amount lost by the donor system is equal to the amount gained by the receptor system. Heat is not a state variable. For his 1947 definition of "heat transfer" for discrete open systems, the author Prigogine carefully explains at some length that his definition of it does not obey a balance law. He describes this as paradoxical.

The situation is clarified by Gyarmati, who shows that his definition of "heat transfer", for continuous-flow systems, really refers not specifically to heat, but rather to transfer of internal energy, as follows. He considers a conceptual small cell in a situation of continuous-flow as a system defined in the so-called Lagrangian way, moving with the local center of mass. The flow of matter across the boundary is zero when considered as a flow of total mass. Nevertheless, if the material constitution is of several chemically distinct components that can diffuse with respect to one another, the system is considered to be open, the diffusive flows of the components being defined with respect to the center of mass of the system, and balancing one another as to mass transfer. Still there can be a distinction between bulk flow of internal energy and diffusive flow of internal energy in this case, because the internal energy density does not have to be constant per unit mass of material, and allowing for non-conservation of internal energy because of local conversion of kinetic energy of bulk flow to internal energy by viscosity.

Gyarmati shows that his definition of "the heat flow vector" is strictly speaking a definition of flow of internal energy, not specifically of heat, and so it turns out that his use here of the word heat is contrary to the strict thermodynamic definition of heat, though it is more or less compatible with historical custom, that often enough did not clearly distinguish between heat and internal energy; he writes "that this relation must be considered to be the exact definition of the concept of heat flow, fairly loosely used in experimental physics and heat technics." Apparently in a different frame of thinking from that of the above-mentioned paradoxical usage in the earlier sections of the historic 1947 work by Prigogine, about discrete systems, this usage of Gyarmati is consistent with the later sections of the same 1947 work by Prigogine, about continuous-flow systems, which use the term "heat flux" in just this way. This usage is also followed by Glansdorff and Prigogine in their 1971 text about continuous-flow systems. They write: "Again the flow of internal energy may be split into a convection flow ρuv and a conduction flow. This conduction flow is by definition the heat flow W. Therefore: j[U] = ρuv + W where u denotes the [internal] energy per unit mass. [These authors actually use the symbols E and e to denote internal energy but their notation has been changed here to accord with the notation of the present article. These authors actually use the symbol U to refer to total energy, including kinetic energy of bulk flow.]" This usage is followed also by other writers on non-equilibrium thermodynamics such as Lebon, Jou, and Casas-Vásquez, and de Groot and Mazur. This usage is described by Bailyn as stating the non-convective flow of internal energy, and is listed as his definition number 1, according to the first law of thermodynamics. This usage is also followed by workers in the kinetic theory of gases. This is not the *ad hoc* definition of "reduced heat flux" of Haase.

In the case of a flowing system of only one chemical constituent, in the Lagrangian representation, there is no distinction between bulk flow and diffusion of matter. Moreover, the flow of matter is zero into or out of the cell that moves with the local center of mass. In effect, in this description, one is dealing with a system effectively closed to the transfer of matter. But still one can validly talk of a distinction between bulk flow and diffusive flow of internal energy, the latter driven by a temperature gradient within the flowing material, and being defined with respect to the local center of mass of the bulk flow. In this case of a virtually closed system, because of the zero matter transfer, as noted above, one can safely distinguish between transfer of energy as work, and transfer of internal energy as heat.

Second Law of Thermodynamics

The second law of thermodynamics states that the total entropy of an isolated system always increases over time, or remains constant in ideal cases where the system is in a steady state or undergoing a reversible process. The increase in entropy accounts for the irreversibility of natural processes, and the asymmetry between future and past.

Historically, the second law was an empirical finding that was accepted as an axiom of thermodynamic theory. Statistical thermodynamics, classical or quantum, explains the microscopic origin of the law.

The second law has been expressed in many ways. Its first formulation is credited to the French scientist Sadi Carnot in 1824, who showed that there is an upper limit to the efficiency of conversion of heat to work in a heat engine.

Introduction

Intuitive Meaning of the Law

The second law is about thermodynamic systems or bodies of matter and radiation, initially each in its own state of internal thermodynamic equilibrium, and separated from one another by walls that partly or wholly allow or prevent the passage of matter and energy between them, or make them mutually inaccessible for their constituents.

The law envisages that the walls are changed by some external agency, making them less restrictive or constraining and more permeable in various ways, and increasing the accessibility, to parts of the overall system, of matter and energy. Thereby a process is defined, establishing new equilibrium states.

The process invariably spreads, disperses, and dissipates matter or energy, or both, amongst the bodies. Some energy, inside or outside the system, is degraded in its ability to do work. This is quantitatively described by increase of entropy. It is the consequence of decrease of constraint by a wall, with a corresponding increase in the accessibility, to the parts of the system, of matter and energy. An increase of constraint by a wall has no effect on an established thermodynamic equilibrium.

For an example of the spreading of matter due to increase of accessibility, one may consider a gas initially confined by an impermeable wall to one of two compartments of an isolated system. The wall is then removed. The gas spreads throughout both compartments. The sum of the entropies of the two compartments increases. Reinsertion of the impermeable wall does not change the spread of the gas between the compartments. For an example of the spreading of energy due to increase of accessibility, one may consider a wall impermeable to matter and energy initially separating two otherwise isolated bodies at different temperatures. A thermodynamic operation makes the wall become permeable only to heat, which then passes from the hotter to the colder body, until their temperatures become equal. The sum of the entropies of the two bodies increases. Restoration of the complete impermeability of the wall does not change the equality of the temperatures. The spreading is a change from heterogeneity towards homogeneity.

It is the unconstraining of the initial equilibrium that causes the increase of entropy and the change towards homogeneity. The following reasoning offers intuitive understanding of this fact. One may imagine that the freshly unconstrained system, still relatively heterogeneous, immediately after the intervention that increased the wall permeability, in its transient condition, arose by spontaneous evolution from an unconstrained previous transient condition of the system. One can then ask, what is the probable such imagined previous condition. The answer is that, overwhelmingly probably, it is just the very same kind of homogeneous condition as that to which the relatively heterogeneous condition will overwhelmingly probably evolve. Obviously, this is possible only in the imagined absence of the constraint that was actually present until its removal. In this light, the reversibility of the dynamics of the evolution of the unconstrained system is evident, in accord with the ordinary laws of microscopic dynamics. It is the removal of the constraint that is effective in causing the change towards homogeneity, not some imagined or apparent "irreversibility" of the laws of spontaneous evolution. This reasoning is of intuitive interest, but is essentially about microstates, and therefore does not belong to macroscopic equilibrium thermodynamics, which studiously ignores consideration of microstates, and non-equilibrium considerations of this kind. It does, however, forestall futile puzzling about some famous proposed "paradoxes", imagining of a "derivation" of an "arrow of time" from the second law, and meaningless speculation about an imagined "low entropy state" of the early universe.

Though it is more or less intuitive to imagine 'spreading', such loose intuition is, for many thermodynamic processes, too vague or imprecise to be usefully quantitatively informative, because competing possibilities of spreading can coexist, for example due to an increase of some constraint combined with decrease of another. The second law justifies the concept of entropy, which makes the notion of 'spreading' suitably precise, allowing quantitative predictions of just how spreading will occur in particular circumstances. It is characteristic of the physical quantity entropy that it refers to states of thermodynamic equilibrium.

General Significance of the Law

The first law of thermodynamics provides the basic definition of internal energy, associated with all thermodynamic systems, and states the rule of conservation of energy. The second law is concerned with the direction of natural processes. It asserts that a natural process runs only in one sense, and is not reversible. For example, heat always flows spontaneously from hotter to colder bodies, and never the reverse, unless external work is performed on the system. Its modern definition is in terms of entropy.

In a fictive reversible process, an infinitesimal increment in the entropy (dS) of a system is defined to result from an infinitesimal transfer of heat (δQ) to a closed system divided by the common temperature (T) of the system and the surroundings which supply the heat:

$$\mathrm{d}S = \frac{\delta Q}{T} \qquad \text{(closed system, idealized fictive reversible process).}$$

The different notations used for infinitesimal amounts of heat (δ) and infinitesimal amounts of entropy (d) is due to that entropy is a function of state, while heat, like work, is not. For an actually possible infinitesimal process without exchange of matter with the surroundings, the second law requires that the increment in system entropy be greater than that:

$$dS > \frac{\delta Q}{T} \quad \text{(closed system, actually possible, irreversible process).}$$

This is because a general process for this case may include work being done on the system by its surroundings, which must have frictional or viscous effects inside the system, and because heat transfer actually occurs only irreversibly, driven by a finite temperature difference.

The zeroth law of thermodynamics in its usual short statement allows recognition that two bodies in a relation of thermal equilibrium have the same temperature, especially that a test body has the same temperature as a reference thermometric body. For a body in thermal equilibrium with another, there are indefinitely many empirical temperature scales, in general respectively depending on the properties of a particular reference thermometric body. The second law allows a distinguished temperature scale, which defines an absolute, thermodynamic temperature, independent of the properties of any particular reference thermometric body.

Various Statements of the Law

The second law of thermodynamics may be expressed in many specific ways, the most prominent classical statements being the statement by Rudolf Clausius (1854), the statement by Lord Kelvin (1851), and the statement in axiomatic thermodynamics by Constantin Carathéodory (1909). These statements cast the law in general physical terms citing the impossibility of certain processes. The Clausius and the Kelvin statements have been shown to be equivalent.

Carnot's Principle

The historical origin of the second law of thermodynamics was in Carnot's principle. It refers to a cycle of a Carnot heat engine, fictively operated in the limiting mode of extreme slowness known as quasi-static, so that the heat and work transfers are between subsystems that are always in their own internal states of thermodynamic equilibrium. The Carnot engine is an idealized device of special interest to engineers who are concerned with the efficiency of heat engines. Carnot's principle was recognized by Carnot at a time when the caloric theory of heat was seriously considered, before the recognition of the first law of thermodynamics, and before the mathematical expression of the concept of entropy. Interpreted in the light of the first law, it is physically equivalent to the second law of thermodynamics, and remains valid today. It states

The efficiency of a quasi-static or reversible Carnot cycle depends only on the temperatures of the two heat reservoirs, and is the same, whatever the working substance. A Carnot engine operated in this way is the most efficient possible heat engine using those two temperatures.

Clausius Statement

The German scientist Rudolf Clausius laid the foundation for the second law of thermodynamics in 1850 by examining the relation between heat transfer and work. His formulation of the second law, which was published in German in 1854, is known as the *Clausius statement*:

Heat can never pass from a colder to a warmer body without some other change, connected therewith, occurring at the same time.

The statement by Clausius uses the concept of 'passage of heat'. As is usual in thermodynamic discussions, this means 'net transfer of energy as heat', and does not refer to contributory transfers one way and the other.

Heat cannot spontaneously flow from cold regions to hot regions without external work being performed on the system, which is evident from ordinary experience of refrigeration, for example. In a refrigerator, heat flows from cold to hot, but only when forced by an external agent, the refrigeration system.

Kelvin Statement

Lord Kelvin expressed the second law as

It is impossible, by means of inanimate material agency, to derive mechanical effect from any portion of matter by cooling it below the temperature of the coldest of the surrounding objects.

Equivalence of the Clausius and the Kelvin Statements

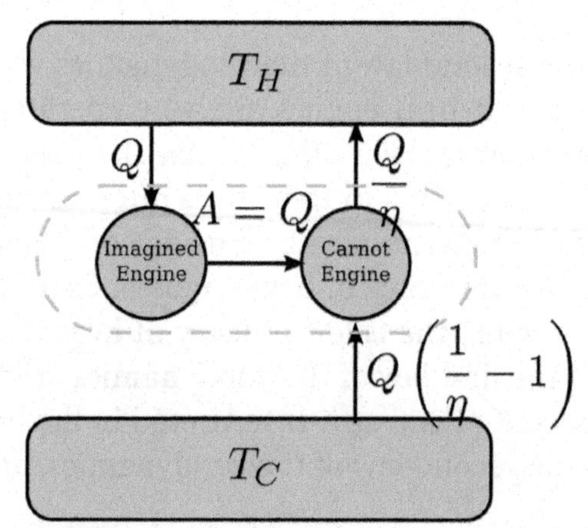

Derive Kelvin Statement from Clausius Statement

Suppose there is an engine violating the Kelvin statement: i.e., one that drains heat and converts it completely into work in a cyclic fashion without any other result. Now pair it with a reversed Carnot engine as shown by the figure. The net and sole effect of this newly created engine consisting of the two engines mentioned is transferring heat $\ddot{A}Q = Q\left(\frac{1}{\eta} - 1\right)$ from the cooler reservoir to the hotter one, which violates the Clausius statement. Thus a violation of the Kelvin statement implies a violation of the Clausius statement, i.e. the Clausius statement implies the Kelvin statement. We can prove in a similar manner that the Kelvin statement implies the Clausius statement, and hence the two are equivalent.

Planck's Proposition

Planck offered the following proposition as derived directly from experience. This is sometimes regarded as his statement of the second law, but he regarded it as a starting point for the derivation of the second law.

> *It is impossible to construct an engine which will work in a complete cycle, and produce no effect except the raising of a weight and cooling of a heat reservoir.*

Relation Between Kelvin's Statement and Planck's Proposition

It is almost customary in textbooks to speak of the "Kelvin-Planck statement" of the law, as for example in the text by ter Haar and Wergeland. One text gives a statement very like Planck's proposition, but attributes it to Kelvin without mention of Planck. One monograph quotes Planck's proposition as the "Kelvin-Planck" formulation, the text naming Kelvin as its author, though it correctly cites Planck in its references. The reader may compare the two statements quoted just above here.

Planck's Statement

Planck stated the second law as follows.

> *Every process occurring in nature proceeds in the sense in which the sum of the entropies of all bodies taking part in the process is increased. In the limit, i.e. for reversible processes, the sum of the entropies remains unchanged.*

Rather like Planck's statement is that of Uhlenbeck and Ford for *irreversible phenomena*.

> ... in an irreversible or spontaneous change from one equilibrium state to another (as for example the equalization of temperature of two bodies A and B, when brought in contact) the entropy always increases.

Principle of Carathéodory

Constantin Carathéodory formulated thermodynamics on a purely mathematical axiomatic foundation. His statement of the second law is known as the Principle of Carathéodory, which may be formulated as follows:

In every neighborhood of any state S of an adiabatically enclosed system there are states inaccessible from S.

With this formulation, he described the concept of adiabatic accessibility for the first time and provided the foundation for a new subfield of classical thermodynamics, often called geometrical thermodynamics. It follows from Carathéodory's principle that quantity of energy quasi-statically transferred as heat is a holonomic process function, in other words, $\delta Q = TdS$.

Though it is almost customary in textbooks to say that Carathéodory's principle expresses the second law and to treat it as equivalent to the Clausius or to the Kelvin-Planck statements, such is not the case. To get all the content of the second law, Carathéodory's principle needs to be supplemented by Planck's principle, that isochoric work always increases the internal energy of a closed system that was initially in its own internal thermodynamic equilibrium.

Planck's Principle

In 1926, Max Planck wrote an important paper on the basics of thermodynamics. He indicated the principle

> The internal energy of a closed system is increased by an adiabatic process, throughout the duration of which, the volume of the system remains constant.

This formulation does not mention heat and does not mention temperature, nor even entropy, and does not necessarily implicitly rely on those concepts, but it implies the content of the second law. A closely related statement is that "Frictional pressure never does positive work." Using a now-obsolete form of words, Planck himself wrote: "The production of heat by friction is irreversible."

Not mentioning entropy, this principle of Planck is stated in physical terms. It is very closely related to the Kelvin statement given just above. It is relevant that for a system at constant volume and mole numbers, the entropy is a monotonic function of the internal energy. Nevertheless, this principle of Planck is not actually Planck's preferred statement of the second law, which is quoted above, in a previous sub-section of the present section of this present article, and relies on the concept of entropy.

A statement that in a sense is complementary to Planck's principle is made by Borgnakke and Sonntag. They do not offer it as a full statement of the second law:

> ... there is only one way in which the entropy of a [closed] system can be decreased, and that is to transfer heat from the system.

Differing from Planck's just foregoing principle, this one is explicitly in terms of entropy change. Of course, removal of matter from a system can also decrease its entropy.

Statement for a System that has a Known Expression of its Internal Energy as a Function of its Extensive State Variables

The second law has been shown to be equivalent to the internal energy U being a weakly convex function, when written as a function of extensive properties (mass, volume, entropy, ...).

Corollaries

Perpetual Motion of the Second Kind

Before the establishment of the Second Law, many people who were interested in inventing a perpetual motion machine had tried to circumvent the restrictions of first law of thermodynamics by extracting the massive internal energy of the environment as the power of the machine. Such a machine is called a "perpetual motion machine of the second kind". The second law declared the impossibility of such machines.

Carnot Theorem

Carnot's theorem (1824) is a principle that limits the maximum efficiency for any possible engine. The efficiency solely depends on the temperature difference between the hot and cold thermal reservoirs. Carnot's theorem states:

- All irreversible heat engines between two heat reservoirs are less efficient than a Carnot engine operating between the same reservoirs.
- All reversible heat engines between two heat reservoirs are equally efficient with a Carnot engine operating between the same reservoirs.

In his ideal model, the heat of caloric converted into work could be reinstated by reversing the motion of the cycle, a concept subsequently known as thermodynamic reversibility. Carnot, however, further postulated that some caloric is lost, not being converted to mechanical work. Hence, no real heat engine could realise the Carnot cycle's reversibility and was condemned to be less efficient.

Though formulated in terms of caloric, rather than entropy, this was an early insight into the second law.

Clausius Inequality

The Clausius theorem (1854) states that in a cyclic process

$$\oint \frac{\delta Q}{T} \leq 0.$$

The equality holds in the reversible case and the '<' is in the irreversible case. The reversible case is used to introduce the state function entropy. This is because in cyclic processes the variation of a state function is zero from state functionality.

Thermodynamic Temperature

For an arbitrary heat engine, the efficiency is:

$$\eta = \frac{W_n}{q_H} = \frac{q_H - q_C}{q_H} = 1 - \frac{q_C}{q_H} \qquad (1)$$

where W_n is for the net work done per cycle. Thus the efficiency depends only on q_C/q_H.

Carnot's theorem states that all reversible engines operating between the same heat reservoirs are equally efficient. Thus, any reversible heat engine operating between temperatures T_1 and T_2 must have the same efficiency, that is to say, the efficiency is the function of temperatures only: $\frac{q_C}{q_H} = f(T_H, T_C)$ (2).

In addition, a reversible heat engine operating between temperatures T_1 and T_3 must have the same efficiency as one consisting of two cycles, one between T_1 and another (intermediate) temperature T_2, and the second between T_2 and T_3. This can only be the case if

$$f(T_1, T_3) = \frac{q_3}{q_1} = \frac{q_2 q_3}{q_1 q_2} = f(T_1, T_2) f(T_2, T_3).$$

Now consider the case where T_1 is a fixed reference temperature: the temperature of the triple point of water. Then for any T_2 and T_3,

$$f(T_2, T_3) = \frac{f(T_1, T_3)}{f(T_1, T_2)} = \frac{273.16 \cdot f(T_1, T_3)}{273.16 \cdot f(T_1, T_2)}.$$

Therefore, if thermodynamic temperature is defined by

$$T = 273.16 \cdot f(T, T)$$

then the function f, viewed as a function of thermodynamic temperature, is simply

$$f(T_2, T_3) = \frac{T_3}{T_2},$$

and the reference temperature T_1 will have the value 273.16. (Of course any reference temperature and any positive numerical value could be used—the choice here corresponds to the Kelvin scale.)

Entropy

According to the Clausius equality, for a reversible process

$$\oint \frac{\delta Q}{T} = 0$$

That means the line integral $\int_L \frac{\delta Q}{T}$ is path independent.

So we can define a state function S called entropy, which satisfies

$$dS = \frac{\delta Q}{T}$$

With this we can only obtain the difference of entropy by integrating the above formula. To obtain the absolute value, we need the Third Law of Thermodynamics, which states that S=0 at absolute zero for perfect crystals.

For any irreversible process, since entropy is a state function, we can always connect the initial and terminal states with an imaginary reversible process and integrating on that path to calculate the difference in entropy.

Now reverse the reversible process and combine it with the said irreversible process. Applying Clausius inequality on this loop,

$$-\Delta S + \int \frac{\delta Q}{T} = \oint \frac{\delta Q}{T} < 0$$

Thus,

$$\Delta S \geq \int \frac{\delta Q}{T}$$

where the equality holds if the transformation is reversible.

Notice that if the process is an adiabatic process, then so .

Energy, Available Useful Work

An important and revealing idealized special case is to consider applying the Second Law to the scenario of an isolated system (called the total system or universe), made up of two parts: a sub-system of interest, and the sub-system's surroundings. These surroundings are imagined to be so large that they can be considered as an *unlimited* heat reservoir at temperature T_R and pressure P_R — so that no matter how much heat is transferred to (or from) the sub-system, the temperature of the surroundings will remain T_R; and no matter how much the volume of the sub-system expands (or contracts), the pressure of the surroundings will remain P_R.

Whatever changes to dS and dS_R occur in the entropies of the sub-system and the surroundings individually, according to the Second Law the entropy S_{tot} of the isolated total system must not decrease:

$$dS_{\text{tot}} = dS + dS_R \geq 0$$

According to the First Law of Thermodynamics, the change dU in the internal energy of the sub-system is the sum of the heat δq added to the sub-system, *less* any work δw done *by* the sub-system, *plus* any net chemical energy entering the sub-system $d\sum\mu_{iR}$-N_i, so that:

$$dU = \delta q - \delta w + d(\sum \mu_{iR} N_i)$$

where μ_{iR} are the chemical potentials of chemical species in the external surroundings.

Now the heat leaving the reservoir and entering the sub-system is

$$\delta q = T_R(-dS_R) \leq T_R dS$$

where we have first used the definition of entropy in classical thermodynamics (alternatively, in statistical thermodynamics, the relation between entropy change, temperature and absorbed heat can be derived); and then the Second Law inequality from above.

It therefore follows that any net work δw done by the sub-system must obey

$$\delta w \leq -dU + T_R dS + \sum \mu_{iR} dN_i$$

It is useful to separate the work δw done by the subsystem into the *useful* work δw_u that can be done *by* the sub-system, over and beyond the work $p_R\, dV$ done merely by the sub-system expanding against the surrounding external pressure, giving the following relation for the useful work (exergy) that can be done:

$$\delta w_u \leq -d(U - T_R S + p_R V - \sum \mu_{iR} N_i)$$

It is convenient to define the right-hand-side as the exact derivative of a thermodynamic potential, called the *availability* or *exergy* *E* of the subsystem,

$$E = U - T_R S + p_R V - \sum \mu_{iR} N_i$$

The Second Law therefore implies that for any process which can be considered as divided simply into a subsystem, and an unlimited temperature and pressure reservoir with which it is in contact,

$$dE + \delta w_u \leq 0$$

i.e. the change in the subsystem's exergy plus the useful work done *by* the subsystem (or, the change in the subsystem's exergy less any work, additional to that done by the pressure reservoir, done *on* the system) must be less than or equal to zero.

In sum, if a proper *infinite-reservoir-like* reference state is chosen as the system surroundings in the real world, then the Second Law predicts a decrease in *E* for an irreversible process and no change for a reversible process.

$$dS_{tot} \geq 0 \text{ Is equivalent to } dE + \delta w_u \leq 0$$

This expression together with the associated reference state permits a design engineer working at the macroscopic scale (above the thermodynamic limit) to utilize the Second Law without directly measuring or considering entropy change in a total isolated system. Those changes have already been considered by the assumption that the system under consideration can reach equilibrium with the reference state without altering the reference state. An efficiency for a process or collection of processes that compares it to the reversible ideal may also be found.

This approach to the Second Law is widely utilized in engineering practice, environmental accounting, systems ecology, and other disciplines.

History

The first theory of the conversion of heat into mechanical work is due to Nicolas Léonard Sadi Carnot in 1824. He was the first to realize correctly that the efficiency of this conversion depends on the difference of temperature between an engine and its environment.

Nicolas Léonard Sadi Carnot in the traditional uniform of a student of the École Polytechnique.

Recognizing the significance of James Prescott Joule's work on the conservation of energy, Rudolf Clausius was the first to formulate the second law during 1850, in this form: heat does not flow *spontaneously* from cold to hot bodies. While common knowledge now, this was contrary to the caloric theory of heat popular at the time, which considered heat as a fluid. From there he was able to infer the principle of Sadi Carnot and the definition of entropy (1865).

Established during the 19th century, the Kelvin-Planck statement of the Second Law says, "It is impossible for any device that operates on a cycle to receive heat from a single reservoir and produce a net amount of work." This was shown to be equivalent to the statement of Clausius.

The ergodic hypothesis is also important for the Boltzmann approach. It says that, over long periods of time, the time spent in some region of the phase space of microstates with the same energy is proportional to the volume of this region, i.e. that all accessible microstates are equally probable over a long period of time. Equivalently, it says that time average and average over the statistical ensemble are the same.

There is a traditional doctrine, starting with Clausius, that entropy can be understood in terms of molecular 'disorder' within a macroscopic system. This doctrine is obsolescent.

Account Given by Clausius

Rudolf Clausius

In 1856, the German physicist Rudolf Clausius stated what he called the "second fundamental theorem in the mechanical theory of heat" in the following form:

$$\int \frac{\delta Q}{T} = -N$$

where Q is heat, T is temperature and N is the "equivalence-value" of all uncompensated transformations involved in a cyclical process. Later, in 1865, Clausius would come to define "equivalence-value" as entropy. On the heels of this definition, that same year, the most famous version of the second law was read in a presentation at the Philosophical Society of Zurich on April 24, in which, in the end of his presentation, Clausius concludes:

The Entropy of the Universe Tends to a Maximum.

This statement is the best-known phrasing of the second law. Because of the looseness of its language, e.g. universe, as well as lack of specific conditions, e.g. open, closed, or isolated, many people take this simple statement to mean that the second law of thermodynamics applies virtually to every subject imaginable. This, of course, is not true; this statement is only a simplified version of a more extended and precise description.

In terms of time variation, the mathematical statement of the second law for an isolated system undergoing an arbitrary transformation is:

$$\frac{dS}{dt} \geq 0$$

where

S is the entropy of the system and

t is time.

The equality sign applies after equilibration. An alternative way of formulating of the second law for isolated systems is:

$$\frac{dS}{dt} = \dot{S}_i \text{ with } \dot{S}_i \geq 0$$

with $\dot{S}_i$ the sum of the rate of entropy production by all processes inside the system. The advantage of this formulation is that it shows the effect of the entropy production. The rate of entropy production is a very important concept since it determines (limits) the efficiency of thermal machines. Multiplied with ambient temperature T_a it gives the so-called dissipated energy $P_{diss} = T_a\dot{S}_i$.

The expression of the second law for closed systems (so, allowing heat exchange and moving boundaries, but not exchange of matter) is:

$$\frac{dS}{dt} = \frac{\dot{Q}}{T} + \dot{S}_i \text{ with } \dot{S}_i \geq 0$$

Here

$\dot{Q}$ is the heat flow into the system

T is the temperature at the point where the heat enters the system.

The equality sign holds in the case that only reversible processes take place inside the system. If irreversible processes take place (which is the case in real systems in operation) the >-sign holds. If heat is supplied to the system at several places we have to take the algebraic sum of the corresponding terms.

For open systems (also allowing exchange of matter):

$$\frac{dS}{dt} = \frac{\dot{Q}}{T} + \dot{S} + \dot{S}_i \text{ with } \dot{S}_i \geq 0$$

Here $\dot{S}$ is the flow of entropy into the system associated with the flow of matter entering the system. It should not be confused with the time derivative of the entropy. If matter is supplied at several places we have to take the algebraic sum of these contributions.

Statistical Mechanics

Statistical mechanics gives an explanation for the second law by postulating that a material is composed of atoms and molecules which are in constant motion. A particular set of positions and velocities for each particle in the system is called a microstate of the system and because of the constant motion, the system is constantly changing its microstate. Statistical mechanics postulates that, in equilibrium, each microstate that the system might be in is equally likely to occur, and when this assumption is made, it leads directly to the conclusion that the second law must hold in a statistical sense. That is, the second law will hold on average, with a statistical variation on the order of $1/\sqrt{N}$ where N is the number of particles in the system. For everyday (macroscopic) situations, the probability that the second law will be violated is practically zero. However, for systems with a small number of particles, thermodynamic parameters, including the entropy, may show significant statistical deviations from that predicted by the second law. Classical thermodynamic theory does not deal with these statistical variations.

Derivation from Statistical Mechanics

Due to Loschmidt's paradox, derivations of the Second Law have to make an assumption regarding the past, namely that the system is uncorrelated at some time in the past; this allows for simple probabilistic treatment. This assumption is usually thought as a boundary condition, and thus the second Law is ultimately a consequence of the initial conditions somewhere in the past, probably at the beginning of the universe (the Big Bang), though other scenarios have also been suggested.

Given these assumptions, in statistical mechanics, the Second Law is not a postulate, rather it is a consequence of the fundamental postulate, also known as the equal prior probability postulate, so long as one is clear that simple probability arguments are applied only to the future, while for the past there are auxiliary sources of information which tell us that it was low entropy.The first part of the second law, which states that the entropy of a thermally isolated system can only increase, is a trivial consequence of the equal prior probability postulate, if we restrict the notion of the entropy to systems in thermal equilibrium. The entropy of an isolated system in thermal equilibrium containing an amount of energy of E is:

$$S = k_B \ln\left[\Omega(E)\right]$$

where Ù (E) is the number of quantum states in a small interval between E and $E+\delta E$. Here δE is a macroscopically small energy interval that is kept fixed. Strictly speaking this means that the entropy depends on the choice of δE. However, in the thermodynamic limit (i.e. in the limit of infinitely large system size), the specific entropy (entropy per unit volume or per unit mass) does not depend on δE.

Suppose we have an isolated system whose macroscopic state is specified by a number of variables. These macroscopic variables can, e.g., refer to the total volume, the positions of pistons in the system, etc. Then Ù will depend on the values of these variables. If a variable is not fixed, (e.g. we do not clamp a piston in a certain position), then because all the accessible states are equally likely in equilibrium, the free variable in equilibrium will be such that Ù is maximized as that is the most probable situation in equilibrium.

If the variable was initially fixed to some value then upon release and when the new equilibrium has been reached, the fact the variable will adjust itself so that Ù is maximized, implies that the entropy will have increased or it will have stayed the same (if the value at which the variable was fixed happened to be the equilibrium value). Suppose we start from an equilibrium situation and we suddenly remove a constraint on a variable. Then right after we do this, there are a number Ω of accessible microstates, but equilibrium has not yet been reached, so the actual probabilities of the system being in some accessible state are not yet equal to the prior probability of $1/\Omega$. We have already seen that in the final equilibrium state, the entropy will have increased or have stayed the same relative to the previous equilibrium state. Boltzmann's H-theorem, however, proves that the quantity H increases monotonically as a function of time during the intermediate out of equilibrium state.

Derivation of the Entropy Change for Reversible Processes

The second part of the Second Law states that the entropy change of a system undergoing a reversible process is given by:

$$dS = \frac{\delta Q}{T}$$

where the temperature is defined as:

$$\frac{1}{k_B T} \equiv \beta \equiv \frac{d\ln\left[\Omega(E)\right]}{dE}$$

See here for the justification for this definition. Suppose that the system has some external parameter, x, that can be changed. In general, the energy eigenstates of the system will depend on x. According to the adiabatic theorem of quantum mechanics, in the limit of an infinitely slow change of the system's Hamiltonian, the system will stay in the same energy eigenstate and thus change its energy according to the change in energy of the energy eigenstate it is in.

The generalized force, X, corresponding to the external variable x is defined such that Xdx is the work performed by the system if x is increased by an amount dx. E.g., if x is the volume, then X is the pressure. The generalized force for a system known to be in energy eigenstate E_r is given by:

$$X = -\frac{dE_r}{dx}$$

Since the system can be in any energy eigenstate within an interval of δE, we define the generalized force for the system as the expectation value of the above expression:

$$X = -\left\langle \frac{dE_r}{dx} \right\rangle$$

To evaluate the average, we partition the $\Omega(E)$ energy eigenstates by counting how many of them have a value for $\frac{dE}{dx}$ within a range between Y and $Y+\delta Y$. Calling this number $\Omega_Y(E)$, we have:

$$\Omega(E)=\sum_Y \Omega_Y(E)$$

The average defining the generalized force can now be written:

$$X=-\frac{1}{\Omega(E)}\sum_Y Y\Omega_Y(E)$$

We can relate this to the derivative of the entropy with respect to x at constant energy E as follows. Suppose we change x to x + dx. Then $\grave{U}(E)$ will change because the energy eigenstates depend on x, causing energy eigenstates to move into or out of the range between E and $E+\delta E$. Let's focus again on the energy eigenstates for which $\frac{dE_r}{dx}$ lies within the range between Y and $Y+\delta Y$. Since these energy eigenstates increase in energy by Y dx, all such energy eigenstates that are in the interval ranging from E – Y dx to E move from below E to above E. There are

$$N_Y(E)=\frac{\Omega_Y(E)}{\delta E}Ydx$$

such energy eigenstates. If $Ydx \le \delta E$,, all these energy eigenstates will move into the range between E and $E+\delta E$ and contribute to an increase in $\grave{U}$. The number of energy eigenstates that move from below $E+\delta E$ to above $E+\delta E$ is, of course, given by $N_Y(E+\delta E)$. The difference

$$N_Y(E)-N_Y(E+\delta E)$$

is thus the net contribution to the increase in $\grave{U}$. Note that if Y dx is larger than δE there will be the energy eigenstates that move from below E to above $E+\delta E$. They are counted in both $N_Y(E)$ and $N_Y(E+\delta E)$, therefore the above expression is also valid in that case.

Expressing the above expression as a derivative with respect to E and summing over Y yields the expression:

$$\left(\frac{\partial \Omega}{\partial x}\right)_E=-\sum_Y Y\left(\frac{\partial \Omega_Y}{\partial E}\right)_x=\left(\frac{\partial(\Omega X)}{\partial E}\right)_x$$

The logarithmic derivative of Ω with respect to x is thus given by:

$$\left(\frac{\partial \ln(\Omega)}{\partial x}\right)_E = \beta X + \left(\frac{\partial X}{\partial E}\right)_x$$

The first term is intensive, i.e. it does not scale with system size. In contrast, the last term scales as the inverse system size and will thus vanishes in the thermodynamic limit. We have thus found that:

$$\left(\frac{\partial S}{\partial x}\right)_E = \frac{X}{T}$$

Combining this with

$$\left(\frac{\partial S}{\partial E}\right)_x = \frac{1}{T}$$

Gives:

$$dS = \left(\frac{\partial S}{\partial E}\right)_x dE + \left(\frac{\partial S}{\partial x}\right)_E dx = \frac{dE}{T} + \frac{X}{T}dx = \frac{\delta Q}{T}$$

Derivation for Systems Described by the Canonical Ensemble

If a system is in thermal contact with a heat bath at some temperature T then, in equilibrium, the probability distribution over the energy eigenvalues are given by the canonical ensemble:

$$P_j = \frac{\exp\left(-\frac{E_j}{k_B T}\right)}{Z}$$

Here Z is a factor that normalizes the sum of all the probabilities to 1, this function is known as the partition function. We now consider an infinitesimal reversible change in the temperature and in the external parameters on which the energy levels depend. It follows from the general formula for the entropy:

$$S = -k_B \sum_j P_j \ln(P_j)$$

that

$$dS = -k_B \sum_j \ln(P_j) dP_j$$

Inserting the formula for P_j for the canonical ensemble in here gives:

$$dS = \frac{1}{T}\sum_j E_j dP_j = \frac{1}{T}\sum_j d(E_j P_j) - \frac{1}{T}\sum_j P_j dE_j = \frac{dE + \delta W}{T} = \frac{\delta Q}{T}$$

Living Organisms

There are two principal ways of formulating thermodynamics, (a) through passages from one state of thermodynamic equilibrium to another, and (b) through cyclic processes, by which the system is left unchanged, while the total entropy of the surroundings is increased. These two ways help to understand the processes of life. This topic is mostly beyond the scope of this present article, but has been considered by several authors, such as Erwin Schrödinger, Léon Brillouin and Isaac Asimov. It is also the topic of current research.

To a fair approximation, living organisms may be considered as examples of (b). Approximately, an animal's physical state cycles by the day, leaving the animal nearly unchanged. Animals take in food, water, and oxygen, and, as a result of metabolism, give out breakdown products and heat. Plants take in radiative energy from the sun, which may be regarded as heat, and carbon dioxide and water. They give out oxygen. In this way they grow. Eventually they die, and their remains rot. This can be regarded as a cyclic process. Overall, the sunlight is from a high temperature source, the sun, and its energy is passed to a lower temperature sink, the soil. This is an increase of entropy of the surroundings of the plant. Thus animals and plants obey the second law of thermodynamics, considered in terms of cyclic processes. Simple concepts of efficiency of heat engines are hardly applicable to this problem because they assume closed systems.

From the thermodynamic viewpoint that considers (a), passages from one equilibrium state to another, only a roughly approximate picture appears, because living organisms are never in states of thermodynamic equilibrium. Living organisms must often be considered as open systems, because they take in nutrients and give out waste products. Thermodynamics of open systems is currently often considered in terms of passages from one state of thermodynamic equilibrium to another, or in terms of flows in the approximation of local thermodynamic equilibrium. The problem for living organisms may be further simplified by the approximation of assuming a steady state with unchanging flows. General principles of entropy production for such approximations are subject to unsettled current debate or research. Nevertheless, ideas derived from this viewpoint on the second law of thermodynamics are enlightening about living creatures.

Gravitational Systems

In systems that do not require for their descriptions the general theory of relativity, bodies always have positive heat capacity, meaning that the temperature rises with energy. Therefore, when energy flows from a high-temperature object to a low-temperature object, the source temperature is decreased while the sink temperature is increased; hence temperature differences tend to diminish over time. This is not always the case for systems in which the gravitational force is important and the general theory of relativity is required. Such systems can spontaneously change towards uneven spread of mass and energy. This applies to the universe in large scale, and consequently it may be difficult or impossible to apply the second law to it. Beyond this, the thermodynamics of systems described by the general theory of relativity is beyond the scope of the present article.

Non-equilibrium States

The theory of classical or equilibrium thermodynamics is idealized. A main postulate or assumption, often not even explicitly stated, is the existence of systems in their own internal states of thermodynamic equilibrium. In general, a region of space containing a physical system at a given time, that may be found in nature, is not in thermodynamic equilibrium, read in the most stringent terms. In looser terms, nothing in the entire universe is or has ever been truly in exact thermodynamic equilibrium.

For purposes of physical analysis, it is often enough convenient to make an assumption of thermodynamic equilibrium. Such an assumption may rely on trial and error for its justification. If the assumption is justified, it can often be very valuable and useful because it makes available the theory of thermodynamics. Elements of the equilibrium assumption are that a system is observed to be unchanging over an indefinitely long time, and that there are so many particles in a system, that its particulate nature can be entirely ignored. Under such an equilibrium assumption, in general, there are no macroscopically detectable fluctuations. There is an exception, the case of critical states, which exhibit to the naked eye the phenomenon of critical opalescence. For laboratory studies of critical states, exceptionally long observation times are needed.

In all cases, the assumption of thermodynamic equilibrium, once made, implies as a consequence that no putative candidate "fluctuation" alters the entropy of the system.

It can easily happen that a physical system exhibits internal macroscopic changes that are fast enough to invalidate the assumption of the constancy of the entropy. Or that a physical system has so few particles that the particulate nature is manifest in observable fluctuations. Then the assumption of thermodynamic equilibrium is to be abandoned. There is no unqualified general definition of entropy for non-equilibrium states.

There are intermediate cases, in which the assumption of local thermodynamic equilibrium is a very good approximation, but strictly speaking it is still an approximation, not theoretically ideal. For non-equilibrium situations in general, it may be useful to consider statistical mechanical definitions of other quantities that may be conveniently called 'entropy', but they should not be confused or conflated with thermodynamic entropy properly defined for the second law. These other quantities indeed belong to statistical mechanics, not to thermodynamics, the primary realm of the second law.

The physics of macroscopically observable fluctuations is beyond the scope of this article.

Arrow of Time

The second law of thermodynamics is a physical law that is not symmetric to reversal of the time direction.

The second law has been proposed to supply an explanation of the difference between moving forward and backwards in time, such as why the cause precedes the effect (the causal arrow of time).

Irreversibility

Irreversibility in thermodynamic processes is a consequence of the asymmetric character of thermodynamic operations, and not of any internally irreversible microscopic properties of the bodies. Thermodynamic operations are macroscopic external interventions imposed on the participating bodies, not derived from their internal properties. There are reputed "paradoxes" that arise from failure to recognize this.

Loschmidt's Paradox

Loschmidt's paradox, also known as the reversibility paradox, is the objection that it should not be possible to deduce an irreversible process from the time-symmetric dynamics that describe the microscopic evolution of a macroscopic system.

In the opinion of Schrödinger, "It is now quite obvious in what manner you have to reformulate the law of entropy—or for that matter, all other irreversible statements—so that they be capable of being derived from reversible models. You must not speak of one isolated system but at least of two, which you may for the moment consider isolated from the rest of the world, but not always from each other." The two systems are isolated from each other by the wall, until it is removed by the thermodynamic operation, as envisaged by the law. The thermodynamic operation is externally imposed, not subject to the reversible microscopic dynamical laws that govern the constituents of the systems. It is the cause of the irreversibility. The statement of the law in this present article complies with Schrödinger's advice. The cause–effect relation is logically prior to the second law, not derived from it.

Poincaré Recurrence Theorem

The Poincaré recurrence theorem considers a theoretical microscopic description of an isolated physical system. This may be considered as a model of a thermodynamic system after a thermodynamic operation has removed an internal wall. The system will, after a sufficiently long time, return to a microscopically defined state very close to the initial one. The Poincaré recurrence time is the length of time elapsed until the return. It is exceedingly long, likely longer than the life of the universe, and depends sensitively on the geometry of the wall that was removed by the thermodynamic operation. The recurrence theorem may be perceived as apparently contradicting the second law of thermodynamics. More obviously, however, it is simply a microscopic model of thermodynamic equilibrium in an isolated system formed by removal of a wall between two systems. For a typical thermodynamical system, the recurrence time is so large (many many times longer than the lifetime of the universe) that, for all practical purposes, one cannot observe the recurrence. One might wish, nevertheless, to imagine that one could wait for the Poincaré recurrence, and then re-insert the wall that was removed by the thermodynamic operation. It is then evident that the appearance of irreversibility is due to the utter unpredictability of the Poincaré recurrence given only that the initial state was one of thermodynamic equilibrium, as is the case in macroscopic thermodynamics. Even if one could wait for it, one has no practical possibility of picking the right instant at which to re-insert the wall. The Poincaré recurrence theorem provides a solution to Loschmidt's paradox. If an isolated thermodynamic system could be monitored over increasingly many multiples of the average Poincaré recurrence time, the thermodynamic behavior of the system would become invariant under time reversal.

James Clerk Maxwell

Maxwell's Demon

James Clerk Maxwell imagined one container divided into two parts, *A* and *B*. Both parts are filled with the same gas at equal temperatures and placed next to each other, separated by a wall. Observing the molecules on both sides, an imaginary demon guards a microscopic trapdoor in the wall. When a faster-than-average molecule from *A* flies towards the trapdoor, the demon opens it, and the molecule will fly from *A* to *B*. The average speed of the molecules in *B* will have increased while in *A* they will have slowed down on average. Since average molecular speed corresponds to temperature, the temperature decreases in *A* and increases in *B*, contrary to the second law of thermodynamics.

One response to this question was suggested in 1929 by Leó Szilárd and later by Léon Brillouin. Szilárd pointed out that a real-life Maxwell's demon would need to have some means of measuring molecular speed, and that the act of acquiring information would require an expenditure of energy.

Maxwell's demon repeatedly alters the permeability of the wall between *A* and *B*. It is therefore performing thermodynamic operations on a microscopic scale, not just observing ordinary spontaneous or natural macroscopic thermodynamic processes.

Quotations

The law that entropy always increases holds, I think, the supreme position among the laws of Nature. If someone points out to you that your pet theory of the universe is in disagreement with Maxwell's equations — then so much the worse for Maxwell's equations. If it is found to be contradicted by observation — well, these experimentalists do bungle things sometimes. But if your theory is found to be against the second law of thermodynamics I can give you no hope; there is nothing for it but to collapse in deepest humiliation.

—Sir Arthur Stanley Eddington, The Nature of the Physical World (1927)

There have been nearly as many formulations of the second law as there have been discussions of it.

—Philosopher / Physicist P.W. Bridgman, (1941)

Clausius is the author of the sibyllic utterance, "The energy of the universe is constant; the entropy of the universe tends to a maximum." The objectives of continuum thermomechanics stop far short of explaining the "universe", but within that theory we may easily derive an explicit statement in some ways reminiscent of Clausius, but referring only to a modest object: an isolated body of finite size.

—Truesdell, C., Muncaster, R.G. (1980). Fundamentals of Maxwell's Kinetic Theory of a Simple Monatomic Gas, Treated as a Branch of Rational Mechanics, Academic Press, New York, ISBN 0-12-701350-4, p.17.

Zeroth Law of Thermodynamics

The zeroth law of thermodynamics states that if two thermodynamic systems are each in thermal equilibrium with a third, then they are in thermal equilibrium with each other.

Two systems are said to be in the relation of thermal equilibrium if they are linked by a wall permeable only to heat and they do not change over time. As a convenience of language, systems are sometimes also said to be in a relation of thermal equilibrium if they are not linked so as to be able to transfer heat to each other, but would not do so if they were connected by a wall permeable only to heat. Thermal equilibrium between two systems is a transitive relation.

The physical meaning of the law was expressed by Maxwell in the words: "All heat is of the same kind". For this reason, another statement of the law is "All diathermal walls are equivalent".

The law is important for the mathematical formulation of thermodynamics, which needs the assertion that the relation of thermal equilibrium is an equivalence relation. This information is needed for a mathematical definition of temperature that will agree with the physical existence of valid thermometers.

Zeroth Law as Equivalence Relation

A thermodynamic system is by definition in its own state of internal thermodynamic equilibrium, that is to say, there is no change in its observable state (i.e. macrostate) over time and no flows occur in it. One precise statement of the zeroth law is that the relation of thermal equilibrium is an equivalence relation on pairs of thermodynamic systems. In other words, the set of all systems each in its own state of internal thermodynamic equilibrium may be divided into subsets in which every system belongs to one and only one subset, and is in thermal equilibrium with every other member of that subset, and is not in thermal equilibrium with a member of any other subset. This means that a unique "tag" can be assigned to every system, and if the "tags" of two systems are the same, they are in thermal equilibrium with each other, and if different, they are not. This property is used to justify the use of empirical temperature as a tagging system. Empirical temperature provides further relations of thermally equilibrated systems, such as order and continuity with regard to "hotness" or "coldness", but these are not implied by the standard statement of the zeroth law.

If it is defined that a thermodynamic system is in thermal equilibrium with itself (i.e., thermal equilibrium is reflexive), then the zeroth law may be stated as follows:

If a body A, *be in thermal equilibrium with two other bodies,* B *and* C, *then* B *and* C *are in thermal equilibrium with one another.*

This statement asserts that thermal equilibrium is a left-Euclidean relation between thermodynamic systems. If we also define that every thermodynamic system is in thermal equilibrium with itself, then thermal equilibrium is also a reflexive relation. Binary relations that are both reflexive and Euclidean are equivalence relations. Thus, again implicitly assuming reflexivity, the zeroth law is therefore often expressed as a right-Euclidean statement:

If two systems are in thermal equilibrium with a third system, then they are in thermal equilibrium with each other.

One consequence of an equivalence relationship is that the equilibrium relationship is symmetric: If A is in thermal equilibrium with B, then B is in thermal equilibrium with A. Thus we may say that two systems are in thermal equilibrium with each other, or that they are in mutual equilibrium. Another consequence of equivalence is that thermal equilibrium is a transitive relationship and is occasionally expressed as such:

If A *is in thermal equilibrium with* B *and if* B *is in thermal equilibrium with* C, *then* A *is in thermal equilibrium with* C .

A reflexive, transitive relationship does not guarantee an equivalence relationship. In order for the above statement to be true, *both* reflexivity *and* symmetry must be implicitly assumed.

It is the Euclidean relationships which apply directly to thermometry. An ideal thermometer is a thermometer which does not measurably change the state of the system it is measuring. Assuming that the unchanging reading of an ideal thermometer is a valid "tagging" system for the equivalence classes of a set of equilibrated thermodynamic systems, then if a thermometer gives the same reading for two systems, those two systems are in thermal equilibrium, and if we thermally connect the two systems, there will be no subsequent change in the state of either one. If the readings are different, then thermally connecting the two systems will cause a change in the states of both systems and when the change is complete, they will both yield the same thermometer reading. The zeroth law provides no information regarding this final reading.

Foundation of Temperature

The zeroth law establishes thermal equilibrium as an equivalence relationship. An equivalence relationship on a set (such as the set of all systems each in its own state of internal thermodynamic equilibrium) divides that set into a collection of distinct subsets ("disjoint subsets") where any member of the set is a member of one and only one such subset. In the case of the zeroth law, these subsets consist of systems which are in mutual equilibrium. This partitioning allows any member of the subset to be uniquely "tagged" with a label identifying the subset to which it belongs. Although the labeling may be quite arbitrary, temperature is just such a labeling process which uses the real number system for tagging. The zeroth law justifies the use of suitable thermodynamic systems as thermometers to provide such a labeling, which yield any number of possible empirical temperature scales, and justifies the use of the second law of thermodynamics to provide an absolute, or thermodynamic temperature scale. Such temperature scales bring additional continuity and ordering (i.e., "hot" and "cold") properties to the concept of temperature.

In the space of thermodynamic parameters, zones of constant temperature form a surface, that provides a natural order of nearby surfaces. One may therefore construct a global temperature function that provides a continuous ordering of states. The dimensionality of a surface of constant temperature is one less than the number of thermodynamic parameters, thus, for an ideal gas described with three thermodynamic parameters P, V and N, it is a two-dimensional surface.

For example, if two systems of ideal gases are in equilibrium, then $P_1V_1/N_1 = P_2V_2/N_2$ where P_i is the pressure in the ith system, V_i is the volume, and N_i is the amount (in moles, or simply the number of atoms) of gas.

The surface $PV/N = const$ defines surfaces of equal thermodynamic temperature, and one may label defining T so that $PV/N = RT$, where R is some constant. These systems can now be used as a thermometer to calibrate other systems. Such systems are known as "ideal gas thermometers".

In a sense, focused on in the zeroth law, there is only one kind of diathermal wall or one kind of heat, as expressed by Maxwell's dictum that "All heat is of the same kind". But in another sense, heat is transferred in different ranks, as expressed by Sommerfeld's dictum "Thermodynamics investigates the conditions that govern the transformation of heat into work. It teaches us to recognize temperature as the measure of the work-value of heat. Heat of higher temperature is richer, is capable of doing more work. Work may be regarded as heat of an infinitely high temperature, as unconditionally available heat." This is why temperature is the particular variable indicated by the zeroth law's statement of equivalence.

Physical Meaning of the Usual Statement of the Zeroth Law

The present article states the zeroth law as it is often summarized in textbooks. Nevertheless, this usual statement perhaps does not explicitly convey the full physical meaning that underlies it. The underlying physical meaning was perhaps first clarified by Maxwell in his 1871 textbook.

In Carathéodory's (1909) theory, it is postulated that there exist walls "permeable only to heat", though heat is not explicitly defined in that paper. This postulate is a physical postulate of existence. It does not, however, as worded just previously, say that there is only one kind of heat. This paper of Carathéodory states as proviso 4 of its account of such walls: "Whenever each of the systems S_1 and S_2 is made to reach equilibrium with a third system S_3 under identical conditions, systems S_1 and S_2 are in mutual equilibrium". It is the function of this statement in the paper, not there labeled as the zeroth law, to provide not only for the existence of transfer of energy other than by work or transfer of matter, but further to provide that such transfer is unique in the sense that there is only one kind of such wall, and one kind of such transfer. This is signaled in the postulate of this paper of Carathéodory that precisely one non-deformation variable is needed to complete the specification of a thermodynamic state, beyond the necessary deformation variables, which are not restricted in number. It is therefore not exactly clear what Carathéodory means when in the introduction of this paper he writes "*It is possible to develop the whole theory without assuming the existence of heat, that is of a quantity that is of a different nature from the normal mechanical quantities.*"

Maxwell (1871) discusses at some length ideas which he summarizes by the words "All heat is of the same kind". Modern theorists sometimes express this idea by postulating the existence of a unique one-dimensional *hotness manifold*, into which every proper temperature scale has a monotonic mapping. This may be expressed by the statement that there is only one kind of temperature, regardless of the variety of scales in which it is expressed. Another modern expression of this idea is that "All diathermal walls are equivalent". This might also be expressed by saying that there is precisely one kind of non-mechanical, non-matter-transferring contact equilibrium between thermodynamic systems.

These ideas may be regarded as helping to clarify the physical meaning of the usual statement of the zeroth law of thermodynamics. It is the opinion of Lieb and Yngvason (1999) that the derivation from statistical mechanics of the law of entropy increase is a goal that has so far eluded the deepest thinkers. Thus the idea remains open to consideration that the existence of heat and temperature are needed as coherent primitive concepts for thermodynamics, as expressed, for example, by Maxwell and Planck. On the other hand, Planck in 1926 clarified how the second law can be stated without reference to heat or temperature, by referring to the irreversible and universal nature of friction in natural thermodynamic processes.

History

According to Arnold Sommerfeld, Ralph H. Fowler invented the title 'the zeroth law of thermodynamics' when he was discussing the 1935 text of Saha and Srivastava. They write on page 1 that "every physical quantity must be measurable in numerical terms". They presume that temperature is a physical quantity and then deduce the statement "If a body A is in temperature equilibrium with two bodies B and C, then B and C themselves will be in temperature equilibrium with each other". They then in a self-standing paragraph italicize as if to state their basic postulate: "*Any of the physical properties of A which change with the application of heat may be observed and utilised for the measurement of temperature.*" They do not themselves here use the term 'zeroth law of thermodynamics'. There are very many statements of these physical ideas in the physics literature long before this text, in very similar language. What was new here was just the label 'zeroth law of thermodynamics'. Fowler, with co-author Edward A. Guggenheim, wrote of the zeroth law as follows:

> ...we introduce the postulate: *If two assemblies are each in thermal equilibrium with a third assembly, they are in thermal equilibrium with each other.*

They then proposed that "it may be shown to follow that the condition for thermal equilibrium between several assemblies is the equality of a certain single-valued function of the thermodynamic states of the assemblies, which may be called the temperature *t*, any one of the assemblies being used as a "thermometer" reading the temperature *t* on a suitable scale. This postulate of the "*Existence of temperature*" could with advantage be known as *the zeroth law of thermodynamics*". The first sentence of this present article is a version of this statement. It is not explicitly evident in the existence statement of Fowler and Guggenheim that temperature refers to a unique attribute of a state of a system, such as is expressed in the idea of the hotness manifold. Also their statement refers explicitly to statistical mechanical assemblies, not explicitly to macroscopic thermodynamically defined systems.

Convective Heat Transfer

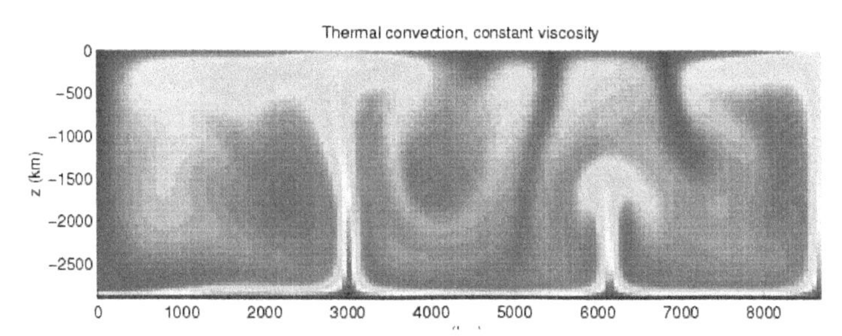

Simulation of thermal convection. Red hues designate hot areas, while regions with blue hues are cold. A hot, less-dense lower boundary layer sends plumes of hot material upwards, and likewise, cold material from the top moves downwards. This illustration is taken from a model of convection in the Earth's mantle.

Convective heat transfer, often referred to simply as convection, is the transfer of heat from one place to another by the movement of fluids. Convection is usually the dominant form of heat transfer(convection) in liquids and gases. Although often discussed as a distinct method of heat transfer, convective heat transfer involves the combined processes of conduction (heat diffusion) and advection (heat transfer by bulk fluid flow).

Convection can be "forced" by movement of a fluid by means other than buoyancy forces (for example, a water pump in an automobile engine). Thermal expansion of fluids may also force convection. In other cases, natural buoyancy forces alone are entirely responsible for fluid motion when the fluid is heated, and this process is called "natural convection". An example is the draft in a chimney or around any fire. In natural convection, an increase in temperature produces a reduction in density, which in turn causes fluid motion due to pressures and forces when fluids of different densities are affected by gravity (or any g-force). For example, when water is heated on a stove, hot water from the bottom of the pan rises, displacing the colder denser liquid, which falls. After heating has stopped, mixing and conduction from this natural convection eventually result in a nearly homogeneous density, and even temperature. Without the presence of gravity (or conditions that cause a g-force of any type), natural convection does not occur, and only forced-convection modes operate. t

The convection heat transfer mode comprises one mechanism. In addition to energy transfer due to specific molecular motion (diffusion), energy is transferred by bulk, or macroscopic, motion of the fluid. This motion is associated with the fact that, at any instant, large numbers of molecules are moving collectively or as aggregates. Such motion, in the presence of a temperature gradient, contributes to heat transfer. Because the molecules in aggregate retain their random motion, the total heat transfer is then due to the superposition of energy transport by random motion of the molecules and by the bulk motion of the fluid. It is customary to use the term convection when referring to this cumulative transport and the term advection when referring to the transport due to bulk fluid motion.

Overview

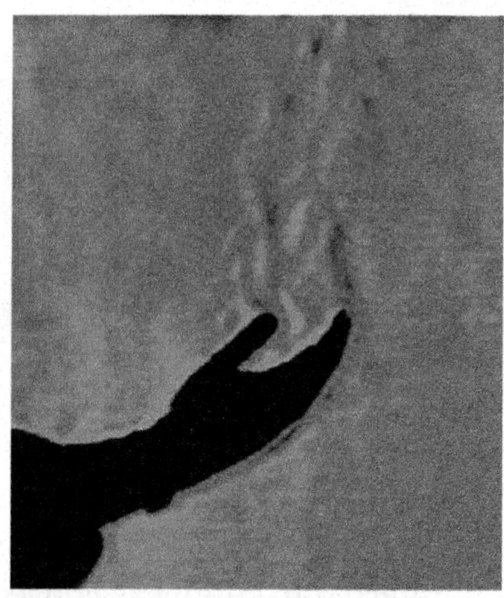

This color schlieren image reveals thermal convection from a human hand (in silhouette form) to the surrounding still atmosphere. Photographed using schlieren equipment.

Convection is the transfer of thermal energy from one place to another by the movement of fluids. Although often discussed as a distinct method of heat transfer, convection describes the combined effects of conduction and fluid flow or mass exchange.

Two types of convective heat transfer may be distinguished:

- Free or natural convection: when fluid motion is caused by buoyancy forces that result from the density variations due to variations of thermal temperature in the fluid. In the absence of an external source, when the fluid is in contact with a hot surface, its molecules separate and scatter, causing the fluid to be less dense. As a consequence, the fluid is displaced while the cooler fluid gets denser and the fluid sinks. Thus, the hotter volume transfers heat towards the cooler volume of that fluid. Familiar examples are the upward flow of air due to a fire or hot object and the circulation of water in a pot that is heated from below.
- Forced convection: when a fluid is forced to flow over the surface by an external source such as fans, by stirring, and pumps, creating an artificially induced convection current.

Internal and external flow can also classify convection. Internal flow occurs when a fluid is enclosed by a solid boundary such when flowing through a pipe. An external flow occurs when a fluid extends indefinitely without encountering a solid surface. Both of these types of convection, either natural or forced, can be internal or external because they are independent of each other.The bulk temperature, or the average fluid temperature, is a convenient reference point for evaluating properties related to convective heat transfer, particularly in applications related to flow in pipes and ducts.

Papers lifted on rising convective air current from warm radiator

Further classification can be made depending on the smoothness and undulations of the solid surfaces. Not all surfaces are smooth, though a bulk of the available information deals with smooth surfaces. Wavy irregular surfaces are commonly encountered in heat transfer devices which include solar collectors, regenerative heat exchangers and underground energy storage systems. They have a significant role to play in the heat transfer processes in these applications. Since they bring in an added complexity due to the undulations in the surfaces, they need to be tackled with mathematical finesse through elegant simplification techniques. Also they do affect the flow and heat transfer characteristics, thereby behaving differently from straight smooth surfaces.

For a visual experience of natural convection, a glass filled with hot water and some red food dye may be placed inside a fish tank with cold, clear water. The convection currents of the red liquid may be seen to rise and fall in different regions, then eventually settle, illustrating the process as heat gradients are dissipated.

Newton's Law of Cooling

Convection-cooling is sometimes loosely assumed to be described by Newton's law of cooling.

Newton's law states that *the rate of heat loss of a body is proportional to the difference in temperatures between the body and its surroundings while under the effects of a breeze*. The constant of proportionality is the heat transfer coefficient. The law applies when the coefficient is independent, or relatively independent, of the temperature difference between object and environment.

In classical natural convective heat transfer, the heat transfer coefficient is dependent on the temperature. However, Newton's law does approximate reality when the temperature changes are relatively small.

Convective Heat Transfer

The basic relationship for heat transfer by convection is:

$$\dot{Q} = hA(T_a - T_b)$$

where $\dot{Q}$ is the heat transferred per unit time, A is the area of the object, h is the heat transfer coefficient, T_a is the object's surface temperature and T_b is the fluid temperature.

The convective heat transfer coefficient is dependent upon the physical properties of the fluid and the physical situation. Values of h have been measured and tabulated for commonly encountered fluids and flow of situations.

NTU Method

The Number of Transfer Units (NTU) Method is used to calculate the rate of heat transfer in heat exchangers (especially counter current exchangers) when there is insufficient information to calculate the Log-Mean Temperature Difference (LMTD). In heat exchanger analysis, if the fluid inlet and outlet temperatures are specified or can be determined by simple energy balance, the LMTD method can be used; but when these temperatures are not available The NTU or The Effectiveness method is used.

To define the effectiveness of a heat exchanger we need to find the maximum possible heat transfer that can be hypothetically achieved in a counter-flow heat exchanger of infinite length. Therefore *one* fluid will experience the maximum possible temperature difference, which is the difference of $T_{h,i} - T_{c,i}$ (The temperature difference between the inlet temperature of the hot stream and the inlet temperature of the cold stream). The method proceeds by calculating the heat capacity rates (i.e. mass flow rate multiplied by specific heat) C_h and C_c for the hot and cold fluids respectively, and denoting the smaller one as C_{min}.

A quantity:

$$q_{max} = C_{min}(T_{h,i} - T_{c,i})$$

is then found,where q_{max} is the maximum heat that could be transferred between the fluids per unit time. C_{min} must be used as it is the fluid with the lowest heat capacity rate that would, in this hypothetical infinite length exchanger, actually undergo the maximum possible temperature change. The other fluid would change temperature more slowly along the heat exchanger length. The method, at this point, is concerned only with the fluid undergoing the maximum temperature change.

The *effectiveness*($\in$), is the ratio between the actual heat transfer rate and the maximum possible heat transfer rate:

$$\in = \frac{q}{q_{max}}$$

where:

$$q = C_h(T_{h,i} - T_{h,o}) = C_c(T_{c,o} - T_{c,i})$$

Effectiveness is dimensionless quantity between 0 and 1. If we know E for a particular heat exchanger, and we know the inlet conditions of the two flow streams we can calculate the amount of heat being transferred between the fluids by:

$$q = \in C_{min}(T_{h,i} - T_{c,i})$$

For any heat exchanger it can be shown that:

$$\in = f(NTU, \frac{C_{min}}{C_{max}})$$

For a given geometry, $\in$ can be calculated using correlations in terms of the "heat capacity ratio"

$$C_r = \frac{C_{min}}{C_{max}}$$

and the *number of transfer units*, NTU

$$NTU = \frac{UA}{C_{min}}$$

where U is the overall heat transfer coefficient and A is the heat transfer area.

For example, the effectiveness of a parallel flow heat exchanger is calculated with:

$$\in = \frac{1 - \exp[-NTU(1 + C_r)]}{1 + C_r}$$

Or the effectiveness of a counter-current flow heat exchanger is calculated with:

$$\in = \frac{1 - \exp[-NTU(1 - C_r)]}{1 - C_r \exp[-NTU(1 - C_r)]}$$

For $C_r = 1$

$$\in = \frac{NTU}{1+NTU}$$

Similar effectiveness relationships can be derived for concentric tube heat exchangers and shell and tube heat exchangers. These relationships are differentiated from one another depending on the type of the flow (counter-current, concurrent, or cross flow), the number of passes (in shell and tube exchangers) and whether a flow stream is mixed or unmixed.

Note that the $C_r = 0$ is a special case in which phase change condensation or evaporation is occurring in the heat exchanger. Hence in this special case the heat exchanger behavior is independent of the flow arrangement. Therefore the effectiveness is given by:

$$\in = 1 - \exp[-NTU]$$

References

- Kittel, C. Kroemer, H. (1980). Thermal Physics, (first edition by Kittel alone 1969), second edition, W. H. Freeman, San Francisco, ISBN 0-7167-1088-9, pp. 49, 227.
- Atkins, P., de Paula, J. (1978/2010). Physical Chemistry, (first edition 1978), ninth edition 2010, Oxford University Press, Oxford UK, ISBN 978-0-19-954337-3, p. 54.
- Aroon Shenoy, Mikhail Sheremet, Ioan Pop, 2016, Convective Flow and Heat Transfer from Wavy Surfaces: Viscous Fluids, Porous Media, and Nanofluids, CRC Press, Taylor & Francis Group, Florida ISBN 978-1-498-76090-4
- "Heat Transfer Mechanisms". Colorado State University. The College of Engineering at Colorado State University. Retrieved 14 September 2015.
- Hawking, SW (1985). "Arrow of time in cosmology". Phys. Rev. D. 32 (10): 2489–2495. Bibcode:1985PhRvD..32.2489H. doi:10.1103/PhysRevD.32.2489. Retrieved 2013-02-15.
- Lebowitz, Joel L. (September 1993). "Boltzmann's Entropy and Time's Arrow" (PDF). Physics Today. 46 (9): 32–38. Bibcode:1993PhT....46i..32L. doi:10.1063/1.881363. Retrieved 2013-02-22.
- Sychev, V. V. (1991). The Differential Equations of Thermodynamics. Taylor & Francis. ISBN 978-1-56032-121-7. Retrieved 2012-11-26.

6

Techniques and Tools of Heat Transfer

Tools and techniques are an important component of any field of study. The following chapter elucidates the various tools and techniques that are related to heat transfer. It serves as a source to understand the major categories of heat transfer, such as heat exchanger, heat spreader, heat sink and loop heat pipe. The aspects elucidated are of vital importance, and provide a better understanding of heat transfer.

Heat Exchanger

A heat exchanger is a device used to transfer heat between one or more fluids. The fluids may be separated by a solid wall to prevent mixing or they may be in direct contact. They are widely used in space heating, refrigeration, air conditioning, power stations, chemical plants, petrochemical plants, petroleum refineries, natural-gas processing, and sewage treatment. The classic example of a heat exchanger is found in an internal combustion engine in which a circulating fluid known as engine coolant flows through radiator coils and air flows past the coils, which cools the coolant and heats the incoming air.

Tubular heat exchanger

Flow Arrangement

There are three primary classifications of heat exchangers according to their flow arrangement. In *parallel-flow* heat exchangers, the two fluids enter the exchanger at the same end, and travel in parallel to one another to the other side. In *counter-flow* heat exchangers the fluids enter the exchanger from opposite ends. The counter current design is the most efficient, in that it can transfer the most heat from the heat (transfer) medium per unit mass due to the fact that the average temperature difference along any unit length is *higher*. See countercurrent exchange. In a *cross-flow* heat exchanger, the fluids travel roughly perpendicular to one another through the exchanger.

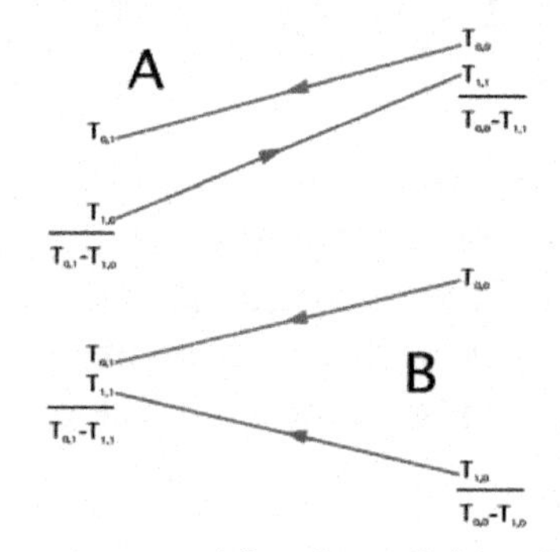

Countercurrent (A) and parallel (B) flows

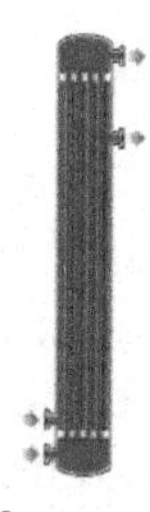

Fig. 1: Shell and tube heat exchanger, single pass (1–1 parallel flow)

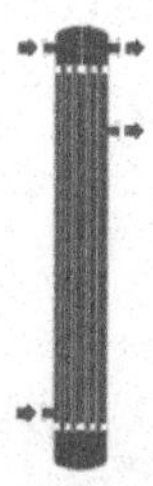

Fig. 2: Shell and tube heat exchanger, 2-pass tube side (1–2 crossflow)

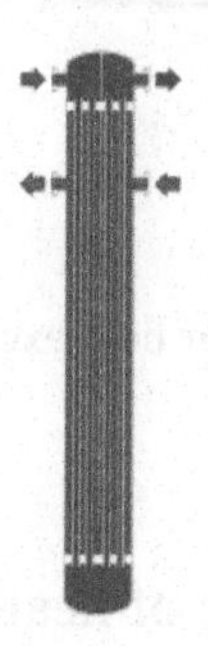

Fig. 3: Shell and tube heat exchanger, 2-pass shell side, 2-pass tube side (2-2 countercurrent)

For efficiency, heat exchangers are designed to maximize the surface area of the wall between the two fluids, while minimizing resistance to fluid flow through the exchanger. The exchanger's performance can also be affected by the addition of fins or corrugations in one or both directions, which increase surface area and may channel fluid flow or induce turbulence.

The driving temperature across the heat transfer surface varies with position, but an appropriate mean temperature can be defined. In most simple systems this is the "log mean temperature difference" (LMTD). Sometimes direct knowledge of the LMTD is not available and the NTU method is used.

Types

Double pipe heat exchangers are the simplest exchangers used in industries. On one hand, these heat exchangers are cheap for both design and maintenance, making them a good choice for small industries. On the other hand, their low efficiency coupled with the high space occupied in large scales, has led modern industries to use more efficient heat exchangers like shell and tube or plate. However, since double pipe heat exchangers are simple, they are used to teach heat exchanger design basics to students as the fundamental rules for all heat exchangers are the same. To start the design of a double pipe heat exchanger, the first step is to calculate the heat duty of the heat exchanger. It must be noted that for easier design, it's better to ignore heat loss to the environment for initial design.

Shell and Tube Heat Exchanger

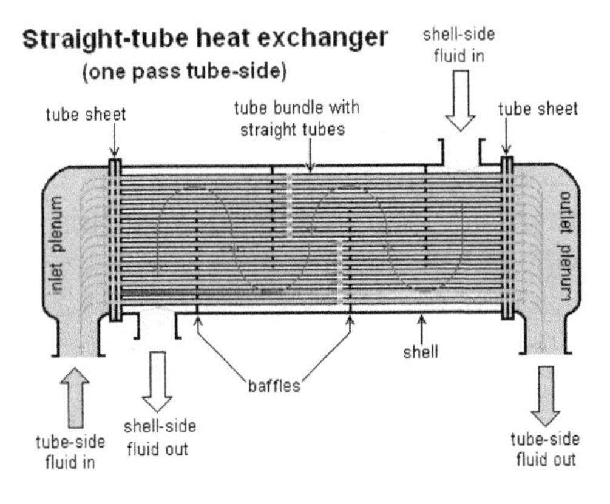

A shell and tube heat exchanger

Shell and tube heat exchangers consist of series of tubes. One set of these tubes contains the fluid that must be either heated or cooled. The second fluid runs over the tubes that are being heated or cooled so that it can either provide the heat or absorb the heat required. A set of tubes is called the tube bundle and can be made up of several types of tubes: plain, longitudinally finned, etc. Shell and tube heat exchangers are typically used for high-pressure applications (with pressures greater than 30 bar and temperatures greater than

260 °C). This is because the shell and tube heat exchangers are robust due to their shape. Several thermal design features must be considered when designing the tubes in the shell and tube heat exchangers: There can be many variations on the shell and tube design. Typically, the ends of each tube are connected to plenums (sometimes called water boxes) through holes in tubesheets. The tubes may be straight or bent in the shape of a U, called U-tubes.

- Tube diameter: Using a small tube diameter makes the heat exchanger both economical and compact. However, it is more likely for the heat exchanger to foul up faster and the small size makes mechanical cleaning of the fouling difficult. To prevail over the fouling and cleaning problems, larger tube diameters can be used. Thus to determine the tube diameter, the available space, cost and fouling nature of the fluids must be considered.
- Tube thickness: The thickness of the wall of the tubes is usually determined to ensure:
  - There is enough room for corrosion
  - That flow-induced vibration has resistance
  - Axial strength
  - Availability of spare parts
  - Hoop strength (to withstand internal tube pressure)
  - Buckling strength (to withstand overpressure in the shell)
- Tube length: heat exchangers are usually cheaper when they have a smaller shell diameter and a long tube length. Thus, typically there is an aim to make the heat exchanger as long as physically possible whilst not exceeding production capabilities. However, there are many limitations for this, including space available at the installation site and the need to ensure tubes are available in lengths that are twice the required length (so they can be withdrawn and replaced). Also, long, thin tubes are difficult to take out and replace.
- Tube pitch: when designing the tubes, it is practical to ensure that the tube pitch (i.e., the centre-centre distance of adjoining tubes) is not less than 1.25 times the tubes' outside diameter. A larger tube pitch leads to a larger overall shell diameter, which leads to a more expensive heat exchanger.
- Tube corrugation: this type of tubes, mainly used for the inner tubes, increases the turbulence of the fluids and the effect is very important in the heat transfer giving a better performance.
- Tube Layout: refers to how tubes are positioned within the shell. There are four main types of tube layout, which are, triangular (30°), rotated triangular (60°),

square (90°) and rotated square (45°). The triangular patterns are employed to give greater heat transfer as they force the fluid to flow in a more turbulent fashion around the piping. Square patterns are employed where high fouling is experienced and cleaning is more regular.

- Baffle Design: baffles are used in shell and tube heat exchangers to direct fluid across the tube bundle. They run perpendicularly to the shell and hold the bundle, preventing the tubes from sagging over a long length. They can also prevent the tubes from vibrating. The most common type of baffle is the segmental baffle. The semicircular segmental baffles are oriented at 180 degrees to the adjacent baffles forcing the fluid to flow upward and downwards between the tube bundle. Baffle spacing is of large thermodynamic concern when designing shell and tube heat exchangers. Baffles must be spaced with consideration for the conversion of pressure drop and heat transfer. For thermo economic optimization it is suggested that the baffles be spaced no closer than 20% of the shell's inner diameter. Having baffles spaced too closely causes a greater pressure drop because of flow redirection. Consequently, having the baffles spaced too far apart means that there may be cooler spots in the corners between baffles. It is also important to ensure the baffles are spaced close enough that the tubes do not sag. The other main type of baffle is the disc and doughnut baffle, which consists of two concentric baffles. An outer, wider baffle looks like a doughnut, whilst the inner baffle is shaped like a disk. This type of baffle forces the fluid to pass around each side of the disk then through the doughnut baffle generating a different type of fluid flow.

Fixed tube liquid-cooled heat exchangers especially suitable for marine and harsh applications can be assembled with brass shells, copper tubes, brass baffles, and forged brass integral end hubs.

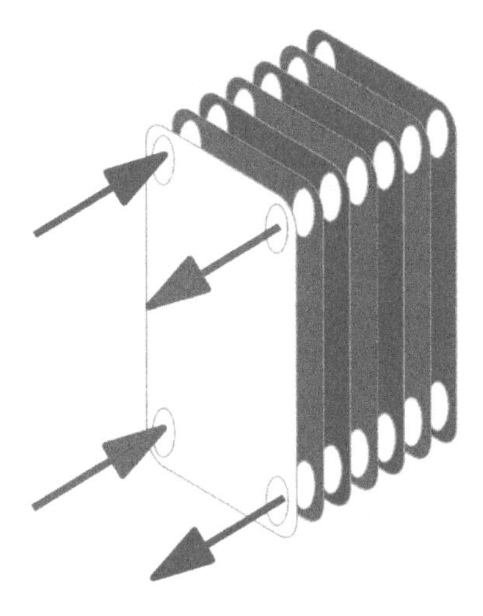

Conceptual diagram of a plate and frame heat exchanger.

A single plate heat exchanger

An interchangeable plate heat exchanger applied to the system of a swimming pool

Plate Heat Exchangers

Another type of heat exchanger is the plate heat exchanger. These exchangers are composed of many thin, slightly separated plates that have very large surface areas and small fluid flow passages for heat transfer. Advances in gasket and brazing technology have made the plate-type heat exchanger increasingly practical. In HVAC applications, large heat exchangers of this type are called *plate-and-frame*; when used in open loops, these heat exchangers are normally of the gasket type to allow periodic disassembly, cleaning, and inspection. There are many types of permanently bonded plate heat exchangers, such as dip-brazed, vacuum-brazed, and welded plate varieties, and they are often specified for closed-loop applications such as refrigeration. Plate heat exchangers also differ in the types of plates that are used, and in the configurations of those plates. Some plates may be stamped with "chevron", dimpled, or other patterns, where others may have machined fins and/or grooves.

When compared to shell and tube exchangers, the stacked-plate arrangement typically has lower volume and cost. Another difference between the two is that plate exchangers typically serve low to medium pressure fluids, compared to medium and high pressures of shell and tube. A third and important difference is that plate exchangers employ more countercurrent flow rather than cross current flow, which allows lower approach temperature differences, high temperature changes, and increased efficiencies.

Plate and Shell Heat Exchanger

A third type of heat exchanger is a plate and shell heat exchanger, which combines plate heat exchanger with shell and tube heat exchanger technologies. The heart of the heat exchanger contains a fully welded circular plate pack made by pressing and cutting round plates and welding them together. Nozzles carry flow in and out of the platepack (the 'Plate side' flowpath). The fully welded platepack is assembled into an outer shell that creates a second flowpath (the 'Shell side'). Plate and shell technology offers high heat transfer, high pressure, high operating temperature, uling and close approach temperature. In particular, it does completely without gaskets, which provides security against leakage at high pressures and temperatures.

Adiabatic Wheel Heat Exchanger

A fourth type of heat exchanger uses an intermediate fluid or solid store to hold heat, which is then moved to the other side of the heat exchanger to be released. Two examples of this are adiabatic wheels, which consist of a large wheel with fine threads rotating through the hot and cold fluids, and fluid heat exchangers.

Plate Fin Heat Exchanger

This type of heat exchanger uses "sandwiched" passages containing fins to increase the effectiveness of the unit. The designs include crossflow and counterflow coupled with various fin configurations such as straight fins, offset fins and wavy fins.

Plate and fin heat exchangers are usually made of aluminum alloys, which provide high heat transfer efficiency. The material cnables the system to operate at a lower temperature difference and reduce the weight of the equipment. Plate and fin heat exchangers are mostly used for low temperature services such as natural gas, helium and oxygen liquefaction plants, air separation plants and transport industries such as motor and aircraft engines.

Advantages of plate and fin heat exchangers:

- High heat transfer efficiency especially in gas treatment
- Larger heat transfer area
- Approximately 5 times lighter in weight than that of shell and tube heat exchanger.
- Able to withstand high pressure

Disadvantages of plate and fin heat exchangers:

- Might cause clogging as the pathways are very narrow
- Difficult to clean the pathways
- Aluminum alloys are susceptible to Mercury Liquid Embrittlement Failure

Pillow Plate Heat Exchanger

A pillow plate exchanger is commonly used in the dairy industry for cooling milk in large direct-expansion stainless steel bulk tanks. The pillow plate allows for cooling across nearly the entire surface area of the tank, without gaps that would occur between pipes welded to the exterior of the tank.

The pillow plate is constructed using a thin sheet of metal spot-welded to the surface of another thicker sheet of metal. The thin plate is welded in a regular pattern of dots or with a serpentine pattern of weld lines. After welding the enclosed space is pressurised with sufficient force to cause the thin metal to bulge out around the welds, providing a space for heat exchanger liquids to flow, and creating a characteristic appearance of a swelled pillow formed out of metal.

Fluid Heat Exchangers

This is a heat exchanger with a gas passing upwards through a shower of fluid (often water), and the fluid is then taken elsewhere before being cooled. This is commonly used for cooling gases whilst also removing certain impurities, thus solving two problems at once. It is widely used in espresso machines as an energy-saving method of cooling super-heated water to use in the extraction of espresso.

Waste Heat Recovery Units

A Waste Heat Recovery Unit (WHRU) is a heat exchanger that recovers heat from a hot gas stream while transferring it to a working medium, typically water or oils. The hot gas stream can be the exhaust gas from a gas turbine or a diesel engine or a waste gas from industry or refinery.

Big systems with high volume and temperature gas streams, typical in industry, can benefit from Steam Rankine Cycle (SRC) in a WHRU, but these cycles are too expensive for small systems. The recovery of heat from low temperature systems requires different working fluids than steam.

An Organic Rankine Cycle (ORC) WHRU can be more efficient at low temperature range using Refrigerant that boil at lower temperatures than water. Typical organic refrigerants are Ammonia, Pentafluoropropane(R-245fa and R-245ca), and Toluene.

The refrigerant is boiled by the heat source in the Evaporator to produce super-heated vapor. This fluid is expanded in the turbine to convert thermal energy to kinetic energy, that is converted to electricity in the electrical generator. This energy transfer process decreases the temperature of the refrigerant that, in turn, condenses. The cycle is closed and completed using a pump to send the fluid back to the evaporator.

Dynamic Scraped Surface Heat Exchanger

Another type of heat exchanger is called "(dynamic) scraped surface heat exchanger". This is mainly used for heating or cooling with high-viscosity products, crystallization processes, evaporation and high-fouling applications. Long running times are achieved due to the continuous scraping of the surface, thus avoiding fouling and achieving a sustainable heat transfer rate during the process.

Phase-change Heat Exchangers

In addition to heating up or cooling down fluids in just a single phase, heat exchangers can be used either to heat a liquid to evaporate (or boil) it or used as condensers to cool a vapor and condense it to a liquid. In chemical plants and refineries, reboilers used to heat incoming feed for distillation towers are often heat exchangers.

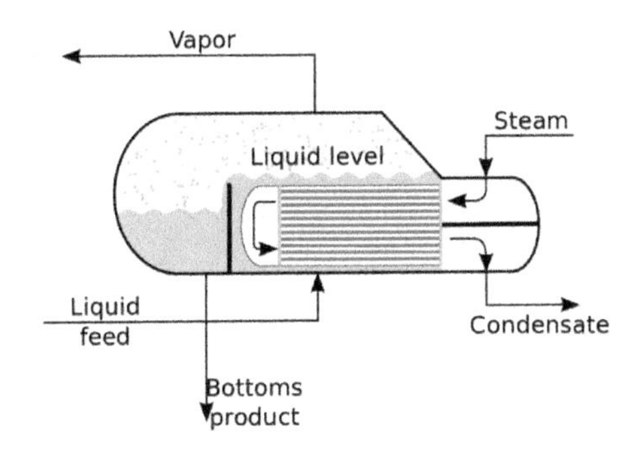

Typical kettle reboiler used for industrial distillation towers

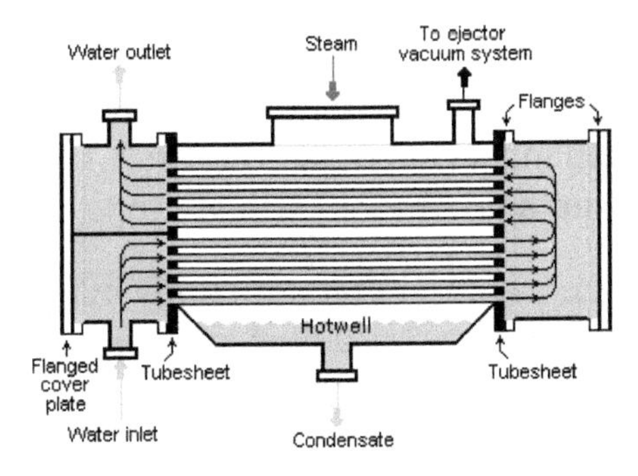

Typical water-cooled surface condenser

Distillation set-ups typically use condensers to condense distillate vapors back into liquid.

Power plants that use steam-driven turbines commonly use heat exchangers to boil water into steam. Heat exchangers or similar units for producing steam from water are often called boilers or steam generators.

In the nuclear power plants called pressurized water reactors, special large heat exchangers pass heat from the primary (reactor plant) system to the secondary (steam plant) system, producing steam from water in the process. These are called steam generators. All fossil-fueled and nuclear power plants using steam-driven turbines have surface condensers to convert the exhaust steam from the turbines into condensate (water) for re-use.

To conserve energy and cooling capacity in chemical and other plants, regenerative heat exchangers can transfer heat from a stream that must be cooled to another stream that must be heated, such as distillate cooling and reboiler feed pre-heating.

This term can also refer to heat exchangers that contain a material within their structure that has a change of phase. This is usually a solid to liquid phase due to the small volume difference between these states. This change of phase effectively acts as a buffer because it occurs at a constant temperature but still allows for the heat exchanger to accept additional heat. One example where this has been investigated is for use in high power aircraft electronics.

Heat exchangers functioning in multiphase flow regimes may be subject to the Ledinegg instability.

Direct Contact Heat Exchangers

Direct contact heat exchangers involve heat transfer between hot and cold streams of two phases in the absence of a separating wall. Thus such heat exchangers can be classified as:

- Gas – liquid
- Immiscible liquid – liquid
- Solid-liquid or solid – gas

Most direct contact heat exchangers fall under the Gas – Liquid category, where heat is transferred between a gas and liquid in the form of drops, films or sprays.

Such types of heat exchangers are used predominantly in air conditioning, humidification, industrial hot water heating, water cooling and condensing plants.

Phases	Continuous phase	Driving force	Change of phase	Examples
Gas – Liquid	Gas	Gravity	No	Spray columns, packed columns
			Yes	Cooling towers, falling droplet evaporators

		Forced	No	Spray coolers/quenchers
		Liquid flow	Yes	Spray condensers/evaporation, jet condensers
	Liquid	Gravity	No	Bubble columns, perforated tray columns
			Yes	Bubble column condensers
		Forced	No	Gas spargers
		Gas flow	Yes	Direct contact evaporators, submerged combustion

Microchannel Heat Exchangers

Micro heat exchangers, Micro-scale heat exchangers, or microstructured heat exchangers are heat exchangers in which (at least one) fluid flows in lateral confinements with typical dimensions below 1 mm. The most typical such confinement are microchannels, which are channels with a hydraulic diameter below 1 mm. Microchannel heat exchangers can be made from metal, ceramic, and even low-cost plastic. Microchannel heat exchangers can be used for many applications including:

- high-performance aircraft gas turbine engines
- heat pumps
- air conditioning
- heat recovery ventilators

Optimization

There are three goals that are normally considered in the optimal design of heat exchangers: (1) Minimizing the pressure drop (pumping power), (2) Maximizing the thermal performance and (3) Minimizing the entropy generation (thermodynamic).

HVAC Air Coils

One of the widest uses of heat exchangers is for air conditioning of buildings and vehicles. This class of heat exchangers is commonly called *air coils*, or just *coils* due to their often-serpentine internal tubing. Liquid-to-air, or air-to-liquid HVAC coils are typically of modified crossflow arrangement. In vehicles, heat coils are often called heater cores.

On the liquid side of these heat exchangers, the common fluids are water, a water-glycol solution, steam, or a refrigerant. For *heating coils*, hot water and steam are the most common, and this heated fluid is supplied by boilers, for example. For *cooling coils*, chilled water and refrigerant are most common. Chilled water is supplied from a chiller that is potentially located very far away, but refrigerant must come from a nearby condensing unit. When a refrigerant is used, the cooling coil is the evaporator in the vapor-compression refrigeration cycle. HVAC coils that use this direct-expansion of refrigerants are commonly called *DX coils*. Some *DX coils* are "microchannel" type.

On the air side of HVAC coils a significant difference exists between those used for heating, and those for cooling. Due to psychrometrics, air that is cooled often has moisture condensing out of it, except with extremely dry air flows. Heating some air increases that airflow's capacity to hold water. So heating coils need not consider moisture condensation on their air-side, but cooling coils *must* be adequately designed and selected to handle their particular *latent* (moisture) as well as the *sensible* (cooling) loads. The water that is removed is called *condensate*.

For many climates, water or steam HVAC coils can be exposed to freezing conditions. Because water expands upon freezing, these somewhat expensive and difficult to replace thin-walled heat exchangers can easily be damaged or destroyed by just one freeze. As such, freeze protection of coils is a major concern of HVAC designers, installers, and operators.

The introduction of indentations placed within the heat exchange fins controlled condensation, allowing water molecules to remain in the cooled air. This invention allowed for refrigeration without icing of the cooling mechanism.

The heat exchangers in direct-combustion furnaces, typical in many residences, are not 'coils'. They are, instead, gas-to-air heat exchangers that are typically made of stamped steel sheet metal. The combustion products pass on one side of these heat exchangers, and air to heat on the other. A *cracked heat exchanger* is therefore a dangerous situation that requires immediate attention because combustion products may enter living space.

Helical-coil Heat Exchangers

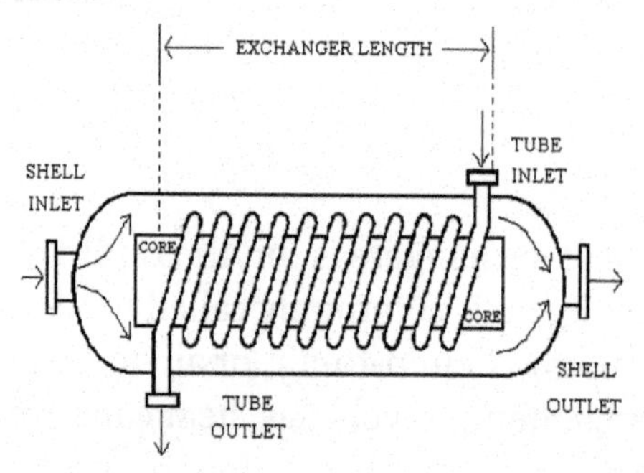

Helical-Coil Heat Exchanger sketch, which consists of a shell, core, and tubes (Scott S. Haraburda design).

Although double-pipe heat exchangers are the simplest to design, the better choice in the following cases would be the helical-coil heat exchanger (HCHE):

- The main advantage of the HCHE, like that for the SHE, is its highly efficient use of space, especially when it's limited and not enough straight pipe can be laid.
- Under conditions of low flowrates (or laminar flow), such that that the typical shell-and-tube exchangers have low heat-transfer coefficients and becoming uneconomical.
- When there is low pressure in one of the fluids, usually from accumulated pressure drops in other process equipment.
- When one of the fluids has components in multiple phases (solids, liquids, and gases), which tends to create mechanical problems during operations, such as plugging of small-diameter tubes. Cleaning of helical coils for these multiple-phase fluids can prove to be more difficult than its shell and tube counterpart; however the helical coil unit would require cleaning less often.

These have been used in the nuclear industry as a method for exchanging heat in a sodium system for large liquid metal fast breeder reactors since the early 1970s, using an HCHE device invented by Charles E. Boardman and John H. Germer. There are several simple methods for designing HCHE for all types of manufacturing industries, such as using the Ramachandra K. Patil (et al.) method from India and the Scott S. Haraburda method from the United States.

However, these are based upon assumptions of estimating inside heat transfer coefficient, predicting flow around the outside of the coil, and upon constant heat flux. Yet, recent experimental data revealed that the empirical correlations are quite in agreement for designing circular and square pattern HCHEs. During studies published in 2015, several researchers found that the boundary conditions of the outer wall of exchangers were essentially constant heat flux conditions in power plant boilers, condensers and evaporators; while convective heat transfer conditions were more appropriate in food, automobile and process industries.

Spiral Heat Exchangers

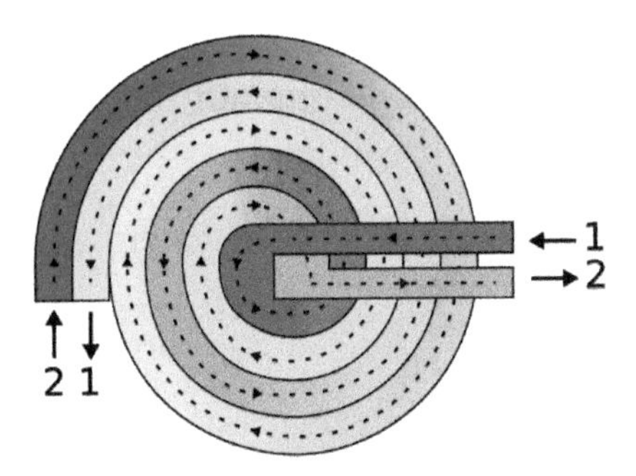

Schematic drawing of a spiral heat exchanger.

A modification to the perpendicular flow of the typical HCHE involves the replacement of shell with another coiled tube, allowing the two fluids to flow parallel to one another, and which requires the use of different design calculations. These are the Spiral Heat Exchangers (SHE), which may refer to a helical (coiled) tube configuration, more generally, the term refers to a pair of flat surfaces that are coiled to form the two channels in a counter-flow arrangement. Each of the two channels has one long curved path. A pair of fluid ports are connected tangentially to the outer arms of the spiral, and axial ports are common, but optional.

The main advantage of the SHE is its highly efficient use of space. This attribute is often leveraged and partially reallocated to gain other improvements in performance, according to well known tradeoffs in heat exchanger design. (A notable tradeoff is capital cost vs operating cost.) A compact SHE may be used to have a smaller footprint and thus lower all-around capital costs, or an over-sized SHE may be used to have less pressure drop, less pumping energy, higher thermal efficiency, and lower energy costs.

Construction

The distance between the sheets in the spiral channels is maintained by using spacer studs that were welded prior to rolling. Once the main spiral pack has been rolled, alternate top and bottom edges are welded and each end closed by a gasketed flat or conical cover bolted to the body. This ensures no mixing of the two fluids occurs. Any leakage is from the periphery cover to the atmosphere, or to a passage that contains the same fluid.

Self cleaning

SHEs are often used in the heating of fluids that contain solids and thus tend to foul the inside of the heat exchanger. The low pressure drop lets the SHE handle fouling more easily. The SHE uses a "self cleaning" mechanism, whereby fouled surfaces cause a localized increase in fluid velocity, thus increasing the drag (or fluid friction) on the fouled surface, thus helping to dislodge the blockage and keep the heat exchanger clean. "The internal walls that make up the heat transfer surface are often rather thick, which makes the SHE very robust, and able to last a long time in demanding environments."They are also easily cleaned, opening out like an oven where any buildup of foulant can be removed by pressure washing.

Self-Cleaning Water filters are used to keep the system clean and running without the need to shut down or replace cartridges and bags.

Flow Arrangements

There are three main types of flows in a spiral heat exchanger:

- Counter-current Flow: Fluids flow in opposite directions. These are used for liquid-liquid, condensing and gas cooling applications. Units are usually mounted vertically when condensing vapour and mounted horizontally when handling high concentrations of solids.
- Spiral Flow/Cross Flow: One fluid is in spiral flow and the other in a cross flow. Spiral flow passages are welded at each side for this type of spiral heat exchanger. This type of flow is suitable for handling low density gas, which passes through the cross flow, avoiding pressure loss. It can be used for liquid-liquid applications if one liquid has a considerably greater flow rate than the other.
- Distributed Vapour/Spiral flow: This design is that of a condenser, and is usually mounted vertically. It is designed to cater for the sub-cooling of both condensate and non-condensables. The coolant moves in a spiral and leaves via the top. Hot gases that enter leave as condensate via the bottom outlet.

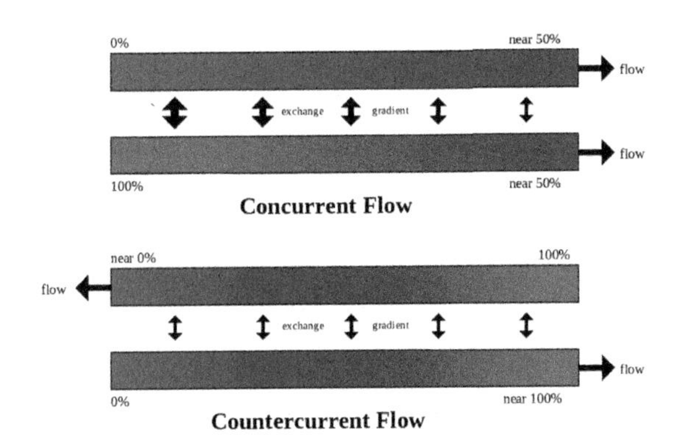

Concurrent and countercurrent flow.

Applications

The SHE is good for applications such as pasteurization, digester heating, heat recovery, pre-heating, and effluent cooling. For sludge treatment, SHEs are generally smaller than other types of heat exchangers.

Selection

Due to the many variables involved, selecting optimal heat exchangers is challenging. Hand calculations are possible, but many iterations are typically needed. As such, heat exchangers are most often selected via computer programs, either by system designers, who are typically engineers, or by equipment vendors.

To select an appropriate heat exchanger, the system designers (or equipment vendors) would firstly consider the design limitations for each heat exchanger type. Though cost is often the primary criterion, several other selection criteria are important:

- High/low pressure limits
- Thermal performance
- Temperature ranges
- Product mix (liquid/liquid, particulates or high-solids liquid)
- Pressure drops across the exchanger
- Fluid flow capacity
- Cleanability, maintenance and repair
- Materials required for construction
- Ability and ease of future expansion
- Material selection, such as copper, aluminum, carbon steel, stainless steel, nickel alloys, ceramic, polymer, and titanium.

Small-diameter coil technologies are becoming more popular in modern air conditioning and refrigeration systems because they have better rates of heat transfer than conventional sized condenser and evaporator coils with round copper tubes and aluminum or copper fin that have been the standard in the HVAC industry. Small diameter coils can withstand the higher pressures required by the new generation of environmentally friendlier refrigerants. Two small diameter coil technologies are currently available for air conditioning and refrigeration products: copper microgroove and brazed aluminum microchannel.

Choosing the right heat exchanger (HX) requires some knowledge of the different heat exchanger types, as well as the environment where the unit must operate. Typically in the manufacturing industry, several differing types of heat exchangers are used for just one process or system to derive the final product. For example, a kettle HX for pre-heating, a double pipe HX for the 'carrier' fluid and a plate and frame HX for final cooling. With sufficient knowledge of heat exchanger types and operating requirements, an appropriate selection can be made to optimise the process.

Monitoring and Maintenance

Online monitoring of commercial heat exchangers is done by tracking the overall heat transfer coefficient. The overall heat transfer coefficient tends to decline over time due to fouling.

$$U=Q/A\Delta T_{lm}$$

By periodically calculating the overall heat transfer coefficient from exchanger flow rates and temperatures, the owner of the heat exchanger can estimate when cleaning the heat exchanger is economically attractive.

Integrity inspection of plate and tubular heat exchanger can be tested in situ by the conductivity or helium gas methods. These methods confirm the integrity of the plates or tubes to prevent any cross contamination and the condition of the gaskets.

Mechanical integrity monitoring of heat exchanger tubes may be conducted through Nondestructive methods such as eddy current testing.

Fouling

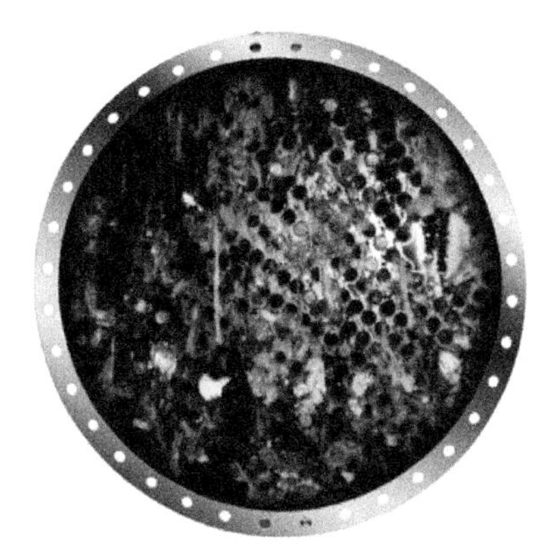

A heat exchanger in a steam power station contaminated with macrofouling.

Fouling occurs when impurities deposit on the heat exchange surface. Deposition of these impurities can decrease heat transfer effectiveness significantly over time and are caused by:

- Low wall shear stress
- Low fluid velocities
- High fluid velocities
- Reaction product solid precipitation
- Precipitation of dissolved impurities due to elevated wall temperatures

The rate of heat exchanger fouling is determined by the rate of particle deposition less re-entrainment/suppression. This model was originally proposed in 1959 by Kern and Seaton.

Crude Oil Exchanger Fouling. In commercial crude oil refining, crude oil is heated from 21 °C (70 °F) to 343 °C (649 °F) prior to entering the distillation column. A series of shell and tube heat exchangers typically exchange heat between crude oil and other oil streams to heat the crude to 260 °C (500 °F) prior to heating in a furnace. Fouling occurs on the crude side of these exchangers due to asphaltene insolubility. The nature of asphaltene solubility in crude oil was successfully modeled by Wiehe and Kennedy. The precipitation of insoluble asphaltenes in crude preheat trains has been successfully modeled as a first order reaction by Ebert and Panchal who expanded on the work of Kern and Seaton.

Cooling Water Fouling. Cooling water systems are susceptible to fouling. Cooling water typically has a high total dissolved solids content and suspended colloidal solids. Localized precipitation of dissolved solids occurs at the heat exchange surface due to wall temperatures higher than bulk fluid temperature. Low fluid velocities (less than 3 ft/s) allow suspended solids to settle on the heat exchange surface. Cooling water is typically on the tube side of a shell and tube exchanger because it's easy to clean. To prevent fouling, designers typically ensure that cooling water velocity is greater than 0.9 m/s and bulk fluid temperature is maintained less than 60 °C (140 °F). Other approaches to control fouling control combine the "blind" application of biocides and anti-scale chemicals with periodic lab testing.

Maintenance

Plate and frame heat exchangers can be disassembled and cleaned periodically. Tubular heat exchangers can be cleaned by such methods as acid cleaning, sandblasting, high-pressure water jet, bullet cleaning, or drill rods.

In large-scale cooling water systems for heat exchangers, water treatment such as purification, addition of chemicals, and testing, is used to minimize fouling of the heat exchange equipment. Other water treatment is also used in steam systems for power plants, etc. to minimize fouling and corrosion of the heat exchange and other equipment.

A variety of companies have started using water borne oscillations technology to prevent biofouling. Without the use of chemicals, this type of technology has helped in providing a low-pressure drop in heat exchangers.

In Nature

Humans

The human nasal passages serve as a heat exchanger, which warms air being inhaled and cools air being exhaled. Its effectiveness can be demonstrated by putting the hand in front of the face and exhaling, first through the nose and then through the mouth. Air exhaled through the nose is substantially cooler. This effect can be enhanced with clothing, by, for example, wearing a scarf over the face while breathing in cold weather.

In species that have external testes (such as humans), the artery to the testis is surrounded by a mesh of veins called the pampiniform plexus. This cools the blood heading to the testis, while reheating the returning blood.

Birds, Fish, Marine Mammals

"Countercurrent" heat exchangers occur naturally in the circulation system of fish, whales and other marine mammals. Arteries to the skin carrying warm blood are intertwined with veins from the skin carrying cold blood, causing the warm arterial blood to exchange heat with the cold venous blood. This reduces the overall heat loss in cold waters. Heat exchangers are also present in the tongue of baleen whales as large volumes of water flow through their mouths. Wading birds use a similar system to limit heat losses from their body through their legs into the water.

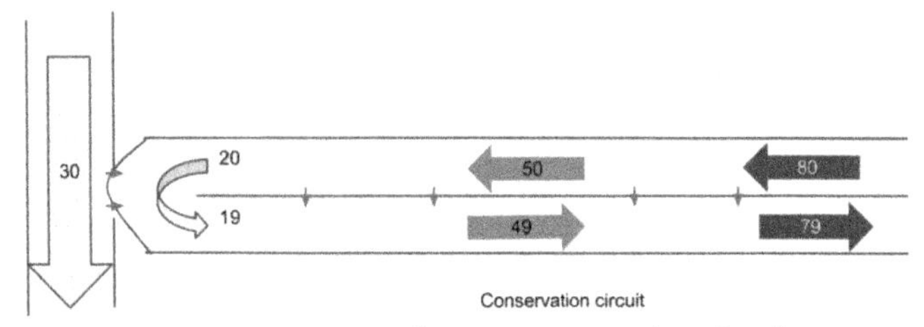

Countercurrent exchange conservation circuit

Carotid Rete

The carotid rete is a counter-current heat exchanging organ in some ungulates. The blood ascending the carotid arteries on its way to the brain, flows via a network of vessels where heat is discharged to the veins of cooler blood descending from the nasal passages. The carotid rete allows Thomson's gazelle to maintain its brain almost 3°C (5.4°F) cooler than the rest of the body, and therefore aids in tolerating bursts in metabolic heat production such as associated with outrunning cheetahs (during which the body temperature exceeds the maximum temperature at which the brain could function).

In Industry

Heat exchangers are widely used in industry both for cooling and heating large scale industrial processes. The type and size of heat exchanger used can be tailored to suit a process depending on the type of fluid, its phase, temperature, density, viscosity, pressures, chemical composition and various other thermodynamic properties.

In many industrial processes there is waste of energy or a heat stream that is being exhausted, heat exchangers can be used to recover this heat and put it to use by heating a different stream in the process. This practice saves a lot of money in industry, as the heat supplied to other streams from the heat exchangers would otherwise come from an external source that is more expensive and more harmful to the environment.

Heat Exchangers are Used in Many Industries, Including:

- Waste water treatment
- Refrigeration
- Wine and beer making
- Petroleum refining
- nuclear power

In waste water treatment, heat exchangers play a vital role in maintaining optimal temperatures within anaerobic digesters to promote the growth of microbes that remove pollutants. Common types of heat exchangers used in this application are the double pipe heat exchanger as well as the plate and frame heat exchanger.

In Aircraft

In commercial aircraft heat exchangers are used to take heat from the engine's oil system to heat cold fuel. This improves fuel efficiency, as well as reduces the possibility of water entrapped in the fuel freezing in components.

Current Market and Forecast

Estimated at US$42.7 billion in 2012, the global demand of heat exchangers will experience robust growth of about 7.8% annually over the next years. The market value is expected to reach US$57.9 billion by 2016 and to approach US$78.16 billion by 2020. Tubular heat exchangers and plate heat exchangers are still the most widely applied product types.

A model of a Simple Heat Exchanger

A simple heat exchange might be thought of as two straight pipes with fluid flow, which are thermally connected. Let the pipes be of equal length *L*, carrying fluids with heat capacity C_i (energy per unit mass per unit change in temperature) and let the mass flow rate of the fluids through the pipes, both in the same direction, be j_i (mass per unit time), where the subscript *i* applies to pipe 1 or pipe 2.

Temperature profiles for the pipes are $T_1(x)$ and $T_2(x)$ where *x* is the distance along the pipe. Assume a steady state, so that the temperature profiles are not functions of time. Assume also that the only transfer of heat from a small volume of fluid in one pipe is to the fluid element in the other pipe at the same position, i.e., there is no transfer of heat along a pipe due to temperature differences in that pipe. By Newton's law of cooling the rate of change in energy of a small volume of fluid is proportional to the difference in temperatures between it and the corresponding element in the other pipe:

$$\frac{du_1}{dt} = \gamma(T_2 - T_1)$$

$$\frac{du_2}{dt} = \gamma(T_1 - T_2)$$

(this is for parallel flow in the same direction and opposite temperature gradients, but for counter-flow heat exchange countercurrent exchange the sign is opposite in the second equation in front of $\gamma(T_1 - T_2)$), where $u_i(x)$ is the thermal energy per unit length and γ is the thermal connection constant per unit length between the two pipes. This change in internal energy results in a change in the temperature of the fluid element. The time rate of change for the fluid element being carried along by the flow is:

$$\frac{du_1}{dt} = J_1 \frac{dT_1}{dx}$$

$$\frac{du_2}{dt} = J_2 \frac{dT_2}{dx}$$

where $J_i = C_i j_i$ is the "thermal mass flow rate". The differential equations governing the heat exchanger may now be written as:

$$J_1 \frac{\partial T_1}{\partial x} = \gamma(T_2 - T_1)$$

$$J_2 \frac{\partial T_2}{\partial x} = \gamma(T_1 - T_2).$$

Note that, since the system is in a steady state, there are no partial derivatives of temperature with respect to time, and since there is no heat transfer along the pipe, there are no second derivatives in x as is found in the heat equation. These two coupled first-order differential equations may be solved to yield:

$$T_1 = A - \frac{Bk_1}{k} e^{-kx}$$

$$T_2 = A + \frac{Bk_2}{k} e^{-kx}$$

where $k_1 = \gamma / J_1$, $k_2 = \gamma / J_2$,

$$k = k_1 + k_2$$

(this is for parallel-flow, but for counter-flow the sign in front of k_2 is negative, so that if $k_2 = k_1$, for the same "thermal mass flow rate" in both opposite directions, the gradient of temperature is constant and the temperatures linear in position x with a constant difference $(T_2 - T_1)$ along the exchanger, explaining why the counter current design countercurrent exchange is the most efficient) and A and B are two as yet undetermined constants of integration. Let T_{10} and T_{20} be the temperatures at x=0 and let T_{1L} and T_{2L} be the temperatures at the end of the pipe at x=L. Define the average temperatures in each pipe as:

$$\overline{T}_1 = \frac{1}{L}\int_0^L T_1(x)dx$$

$$\overline{T}_2 = \frac{1}{L}\int_0^L T_2(x)dx.$$

Using the solutions above, these temperatures are:

$T_{10} = A - \frac{Bk_1}{k}$	$T_{20} = A + \frac{Bk_2}{k}$
$T_{1L} = A - \frac{Bk_1}{k} e^{-kL}$	$T_{2L} = A + \frac{Bk_2}{k} e^{-kL}$
$\overline{T}_1 = A - \frac{Bk_1}{k^2 L}(1 - e^{-kL})$	$\overline{T}_2 = A + \frac{Bk_2}{k^2 L}(1 - e^{-kL}).$

Choosing any two of the temperatures above eliminates the constants of integration, letting us find the other four temperatures. We find the total energy transferred by integrating the expressions for the time rate of change of internal energy per unit length:

$$\frac{dU_1}{dt} = \int_0^L \frac{du_1}{dt} dx = J_1(T_{1L} - T_{10}) = \gamma L(\overline{T}_2 - \overline{T}_1)$$

$$\frac{dU_2}{dt} = \int_0^L \frac{du_2}{dt} dx = J_2(T_{2L} - T_{20}) = \gamma L(\overline{T}_1 - \overline{T}_2).$$

By the conservation of energy, the sum of the two energies is zero. The quantity $\overline{T}_2 - \overline{T}_1$ is known as the *Log mean temperature difference*, and is a measure of the effectiveness of the heat exchanger in transferring heat energy.

Heat Spreader

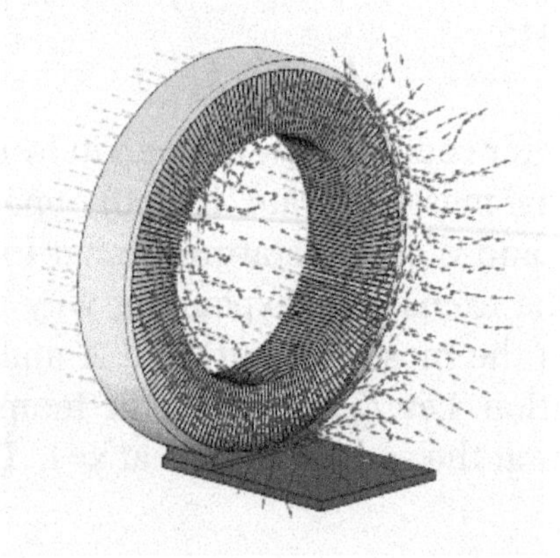

This 120 mm-diameter vapor chamber (heat spreader) heat sink design thermal animation, was created using high resolution CFD analysis, and shows temperature contoured heat sink surface and fluid flow trajectories, predicted using a CFD analysis package, courtesy of NCI.

A heat spreader is a heat exchanger that moves heat between a heat source and a secondary heat exchanger whose surface area and geometry are more favorable than the source. Such a spreader is most often simply a plate made of copper, which has a high thermal conductivity. By definition, heat is "spread out" over this geometry, so that the secondary heat exchanger may be more fully utilized. This has the potential to increase the heat capacity of the total assembly, but the presence of the additional thermal junction will limit total thermal capacity. The high conduction properties of the spreader will make it more effective to function as an air heat exchanger, as opposed to the original (presumably smaller) source. The low heat conduction of air in convection is matched by the higher surface area of the spreader, and heat is radiated more effectively.

A heat spreader is generally used when the heat source tends to have a high heat-flux density, (high heat flow per unit area), and for whatever reason, heat can not be conducted away effectively by the secondary heat exchanger. For instance, this may be because it is air-cooled, giving it a lower heat transfer coefficient than if it were liquid-cooled. A high enough heat exchanger transfer coefficient is sufficient to avoid the need for a heat spreader.

The use of a heat spreader is an important part of an economically optimal design for transferring heat from high to low heat flux media. Examples include:

- A copper-clad bottom on a stove-top cooking container made of steel or stainless steel
- Air-cooling integrated circuits such as a microprocessor
- Air-cooling a photovoltaic cell in a concentrated photovoltaics system

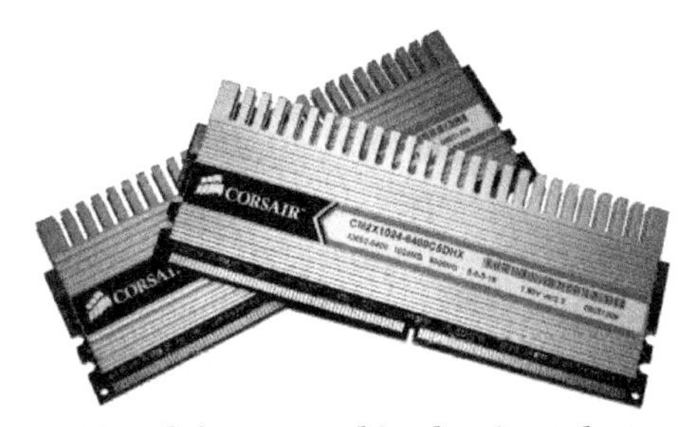

Two memory modules encased in aluminum heat spreaders

Side-by-side comparison of AMD (center) and Intel (sides) integrated heatspreaders common on their microprocessors.

Heat Sink

A fan-cooled heat sink on the processor of a personal computer. To the right is a smaller heat sink cooling another integrated circuit of the motherboard.

A heat sink (also commonly spelled *heatsink*) is a passive heat exchanger that transfers the heat generated by an electronic or a mechanical device to a fluid medium, often air or a liquid coolant, where it is dissipated away from the device, thereby allowing regulation of the device's temperature at optimal levels. In computers, heat sinks are used to cool central processing units or graphics processors. Heat sinks are used with high-power semiconductor devices such as power transistors and optoelectronics such as lasers and light emitting diodes (LEDs), where the heat dissipation ability of the component itself is insufficient to moderate its temperature.

A heat sink is designed to maximize its surface area in contact with the cooling medium surrounding it, such as the air. Air velocity, choice of material, protrusion design and surface treatment are factors that affect the performance of a heat sink. Heat sink attachment methods and thermal interface materials also affect the die temperature of the integrated circuit. Thermal adhesive or thermal grease improve the heat sink's performance by filling air gaps between the heat sink and the heat spreader on the device. A heat sink is usually made out of copper and/or aluminium. Copper is used because it has many desirable properties for thermally efficient and durable heat exchangers. First and foremost, copper is an excellent conductor of heat. This means that copper's high thermal conductivity allows heat to pass through it quickly. Aluminum is used in applications where weight is a big concern.

Heat Transfer Principle

A heat sink transfers thermal energy from a higher temperature device to a lower temperature fluid medium. The fluid medium is frequently air, but can also be water, refrigerants or oil. If the fluid medium is water, the heat sink is frequently called a cold plate. In thermodynamics a heat sink is a heat reservoir that can absorb an arbitrary amount of heat without significantly changing temperature. Practical heat sinks for electronic devices must have a temperature higher than the surroundings to transfer heat by convection, radiation, and conduction. The power supplies of electronics are not 100% efficient, so extra heat is produced that may be detrimental to the function of the device. As such, a heat sink is included in the design to disperse heat to improve efficient energy use.

To understand the principle of a heat sink, consider Fourier's law of heat conduction. Fourier's law of heat conduction, simplified to a one-dimensional form in the x-direction, shows that when there is a temperature gradient in a body, heat will be transferred from the higher temperature region to the lower temperature region. The rate at which heat is transferred by conduction, q_k, is proportional to the product of the temperature gradient and the cross-sectional area through which heat is transferred.

$$q_k = -kA\frac{dT}{dx}$$

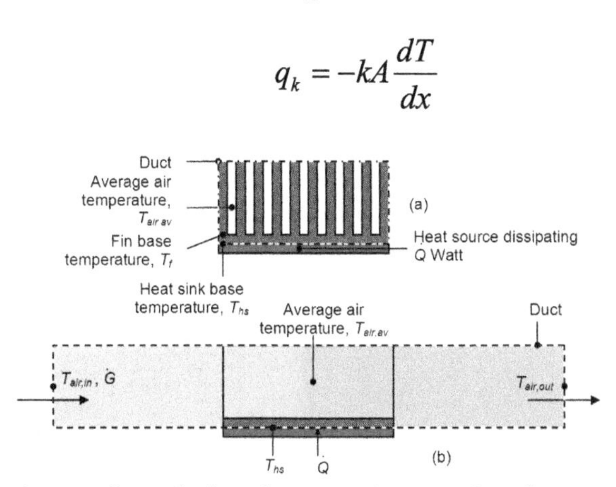

Sketch of a heat sink in a duct used to calculate the governing equations from conservation of energy and Newton's law of cooling.

Consider a heat sink in a duct, where air flows through the duct. It is assumed that the heat sink base is higher in temperature than the air. Applying the conservation of energy, for steady-state conditions, and Newton's law of cooling to the temperature nodes shown in the diagram gives the following set of equations:

$$\dot{Q} = \dot{m}c_{p,in}(T_{air,out} - T_{air,in}) \ (1)$$

$$\frac{T_{hs} \quad T_{air\ av}}{hs} \ (2)$$

where

$$T_{air,av} = \frac{T_{air,in} + T_{air,out}}{2} \ (3)$$

Using the mean air temperature is an assumption that is valid for relatively short heat sinks. When compact heat exchangers are calculated, the logarithmic mean air temperature is used. is the air mass flow rate in kg/s.

The above equations show that

- When the air flow through the heat sink decreases, this results in an increase in the average air temperature. This in turn increases the heat sink base temperature. And additionally, the thermal resistance of the heat sink will also increase. The net result is a higher heat sink base temperature.

  - The increase in heat sink thermal resistance with decrease in flow rate will be shown later in this article.

- The inlet air temperature relates strongly with the heat sink base temperature. For example, if there is recirculation of air in a product, the inlet air temperature is not the ambient air temperature. The inlet air temperature of the heat sink is therefore higher, which also results in a higher heat sink base temperature.
- If there is no air flow around the heat sink, energy cannot be transferred.
- A heat sink is not a device with the "magical ability to absorb heat like a sponge and send it off to a parallel universe".

Natural convection requires free flow of air over the heat sink. If fins are not aligned vertically, or if fins are too close together to allow sufficient air flow between them, the efficiency of the heat sink will decline.

Design Factors

Thermal Resistance

For semiconductor devices used in a variety of consumer and industrial electronics, the idea of *thermal resistance* simplifies the selection of heat sinks. The heat flow between the semiconductor die and ambient air is modeled as a series of resistances to heat flow; there is a resistance from the die to the device case, from the case to the heat sink, and from the heat sink to the ambient air. The sum of these resistances is the total thermal resistance from the die to the ambient air. Thermal resistance is defined as temperature rise per unit of power, analogous to electrical resistance, and is expressed in units of degrees Celsius per watt (°C/W). If the device dissipation in watts is known, and the total thermal resistance is calculated, the temperature rise of the die over the ambient air can be calculated.

The idea of thermal resistance of a semiconductor heat sink is an approximation. It does not take into account non-uniform distribution of heat over a device or heat sink. It only models a system in thermal equilibrium, and does not take into account the change in temperatures with time. Nor does it reflect the non-linearity of radiation and convection with respect to temperature rise. However, manufacturers tabulate typical values of thermal resistance for heat sinks and semiconductor devices, which allows selection of commercially manufactured heat sinks to be simplified.

Commercial extruded aluminium heat sinks have a thermal resistance (heat sink to ambient air) ranging from 0.4 °C/W for a large sink meant for TO3 devices, up to as high as 85 °C/W for a clip-on heat sink for a TO92 small plastic case. The popular

2N3055 power transistor in a TO3 case has an internal thermal resistance from junction to case of 1.52 °C/W. The contact between the device case and heat sink may have a thermal resistance of between 0.5 up to 1.7 °C/W, depending on the case size, and use of grease or insulating mica washer.

Material

The most common heat sink materials are aluminium alloys. Aluminium alloy 1050A has one of the higher thermal conductivity values at 229 W/m•K but is mechanically soft. Aluminium alloys 6060 (low stress), 6061 and 6063 are commonly used, with thermal conductivity values of 166 and 201 W/m•K, respectively. The values depend on the temper of the alloy.

Copper has excellent heat sink properties in terms of its thermal conductivity, corrosion resistance, biofouling resistance, and antimicrobial resistance. Copper has around twice the thermal conductivity of alu-minium and faster, more efficient heat absorption. Its main applications are in indus-trial facilities, power plants, solar thermal water systems, HVAC systems, gas water heaters, forced air heating and cooling systems, geothermal heating and cooling, and electronic systems.

Copper is three times as dense and more expensive than aluminium. Copper heat sinks are machined and skived. Another method of manufacture is to solder the fins into the heat sink base. Aluminium heat sinks can be extruded, but the less ductile copper cannot.

Diamond is another heat sink material, and its thermal conductivity of 2000 W/m•K exceeds copper five-fold.In contrast to metals, where heat is conducted by delocalized electrons, lattice vibrations are responsible for diamond's very high thermal conductivity. For thermal management applications, the outstanding thermal conductivity and diffusivity of diamond is an essential. Nowadays synthetic diamond is used as submounts for high-power integrated circuits and laser diodes.

Composite materials can be used. Examples are a copper-tungsten pseudoalloy, AlSiC (silicon carbide in aluminium matrix), Dymalloy (diamond in copper-silver alloy matrix), and E-Material (beryllium oxide in beryllium matrix). Such materials are often used as substrates for chips, as their thermal expansion coefficient can be matched to ceramics and semiconductors.

Fin Efficiency

Fin efficiency is one of the parameters which makes a higher thermal conductivity material important. A fin of a heat sink may be considered to be a flat plate with heat flowing in one end and being dissipated into the surrounding fluid as it travels to the other. As heat flows through the fin, the combination of the thermal resistance of the heat sink impeding the flow and the heat lost due to convection, the temperature of the fin and, therefore, the heat transfer to the fluid, will decrease from the base to the end of the fin. Fin efficiency is defined as the actual heat transferred by the fin, divided by the heat transfer were the fin to be isothermal (hypothetically the fin having infinite thermal conductivity). Equations 6 and 7 are applicable for straight fins.

$$\eta_f = \frac{\tanh(mL_c)}{mL_c} \quad (6)$$

$$mL_c = \sqrt{\frac{2h_f}{kt_f}} L_f \quad (7)$$

Where:

- h_f is the convection coefficient of the fin
    - o Air: 10 to 100 W/(m²K)
    - o Water: 500 to 10,000 W/(m²K)
- k is the thermal conductivity of the fin material
    - o Aluminium: 120 to 240 W/(m·K)
- L_f is the fin height (m)
- t_f is the fin thickness (m)

Fin efficiency is increased by decreasing the fin aspect ratio (making them thicker or shorter), or by using more conductive material (copper instead of aluminium, for example).

Spreading Resistance

Another parameter that concerns the thermal conductivity of the heat sink material is spreading resistance. Spreading resistance occurs when thermal energy is transferred from a small area to a larger area in a substance with finite thermal conductivity. In a heat sink, this means that heat does not distribute uniformly through the heat sink base. The spreading resistance phenomenon is shown by how the heat travels from the heat source location and causes a large temperature gradient between the heat source and the edges of the heat sink. This means that some fins are at a lower temperature than if the heat source were uniform across the base of the heat sink. This nonuniformity increases the heat sink's effective thermal resistance.

To decrease the spreading resistance in the base of a heat sink:

- Increase the base thickness

- Choose a different material with better thermal conductivity
- Use a vapor chamber or heat pipe in the heat sink base.

Optimization (shape of the fins, location of fins, ...)

The shape of fins must be optimized to maximize heat transfer density and to minimize the pressure drop in the coolant fluid across the heat sink where the space and the materials used for the finned surfaces are constraints. Many shapes exist in the literature ranging from elliptical and cylindrical cylinders, conical shapes to more sharpened surfaces such as rhombus, square sections.

Fin Arrangements

 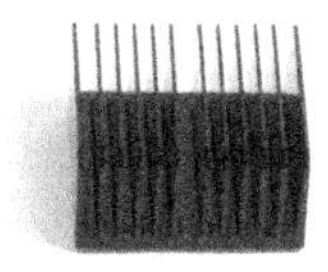 

A pin, straight and flared fin heat sink types

A pin fin heat sink is a heat sink that has pins that extend from its base. The pins can be cylindrical, elliptical or square. A pin is by far one of the more common heat sink types available on the market.A second type of heat sink fin arrangement is the straight fin. These run the entire length of the heat sink. A variation on the straight fin heat sink is a cross cut heat sink. A straight fin heat sink is cut at regular intervals.

In general, the more surface area a heat sink has, the better it works. However, this is not always true. The concept of a pin fin heat sink is to try to pack as much surface area into a given volume as possible. As well, it works well in any orientation. Kordyban has compared the performance of a pin fin and a straight fin heat sink of similar dimensions. Although the pin fin has 194 cm² surface area while the straight fin has 58 cm², the temperature difference between the heat sink base and the ambient air for the pin fin is 50 °C. For the straight fin it was 44 °C or 6 °C better than the pin fin. Pin fin heat sink performance is significantly better than straight fins when used in their intended application where the fluid flows axially along the pins rather than only tangentially across the pins.

Comparison of a pin fin and straight fin heat sink of similar dimensions. Adapted from data

Heat sink fin type	Width [cm]	Length [cm]	Height [cm]	Surface area [cm²]	Volume [cm³]	Temperature difference, $T_{case} - T_{air}$ [°C]
Straight	2.5	2.5	3.2	58	20	44
Pin	3.8	3.8	1.7	194	24	51

Another configuration is the flared fin heat sink; its fins are not parallel to each other, as shown in figure 5. Flaring the fins decreases flow resistance and makes more air go through the heat sink fin channel; otherwise, more air would bypass the fins. Slanting them keeps the overall dimensions the same, but offers longer fins. Forghan, et al. have published data on tests conducted on pin fin, straight fin and flared fin heat sinks. They found that for low approach air velocity, typically around 1 m/s, the thermal performance is at least 20% better than straight fin heat sinks. Lasance and Eggink also found that for the bypass configurations that they tested, the flared heat sink performed better than the other heat sinks tested.

High Conductivity Materials

In the recent years, utilization of high-conductivity materials (inserts) have been proposed for electronic cooling and for enhancing the heat removal from small chips to a heat sink. Because the space occupied by high conductivity materials together with the cost are the two elements of major concern. Therefore, seeking for more efficient designs of high conductivity pathways, embedded into a heat generating body constitutes a formidable challenge.

Cavities (inverted fins)

Cavities (inverted fins) embedded in a heat source are the regions formed between adjacent fins that stand for the essential promoters of nucleate boiling or condensation. These cavities are usually utilized to extract heat from a variety of heat generating bodies to a heat sink.

Conductive Thick Plate Between the Heat Source and the Heat Sink

Placing a conductive thick plate as a heat transfer interface between a heat source and a cold flowing fluid (or any other heat sink) may improve the cooling performance. In such arrangement, the heat source is cooled under the thick plate instead of being cooled in direct contact with the cooling fluid. It is shown that the thick plate can significantly improve the heat transfer between the heat source and the cooling fluid by way of conducting the heat current in an optimal manner. The two most attractive advantages of this method are that no additional pumping power and no extra heat transfer surface area, that is quite different from fins (extended surfaces).

Surface Color

The heat transfer from the heat sink occurs by convection of the surrounding air, conduction through the air, and radiation.

Heat transfer by radiation is a function of both the heat sink temperature, and the temperature of the surroundings that the heat sink is optically coupled with. When both of these temperatures are on the order of 0 °C to 100 °C, the contribution of radiation compared to convection is generally small, and this factor is often neglected. In this case, finned heat sinks operating in either natural-convection or forced-flow will not be affected significantly by surface emissivity.

A server grade flash memory card with a black heat sink.

In situations where convection is low, such as a flat non-finned panel with low airflow, radiative cooling can be a significant factor. Here the surface properties may be an important design factor. Matte-black surfaces will radiate much more efficiently than shiny bare metal. A shiny metal surface has low emissivity. The emissivity of a material is tremendously frequency dependent, and is related to absorptivity (of which shiny metal surfaces have very little). For most materials, the emissivity in the visible spectrum is similar to the emissivity in the infrared spectrum; however there are exceptions, notably certain metal oxides that are used as "selective surfaces".

In a vacuum or in outer space, there is no convective heat transfer, thus in these environments, radiation is the only factor governing heat flow between the heat sink and the environment. For a satellite in space, a 100 °C (373 Kelvin) surface facing the sun will absorb a lot of radiant heat, because the sun's surface temperature is nearly 6000 Kelvin, whereas the same surface facing deep-space will radiate a lot of heat, since deep-space has an effective temperature of only a few Kelvin.

Engineering Applications

Microprocessor Cooling

Heat dissipation is an unavoidable by-product of electronic devices and circuits. In general, the temperature of the device or component will depend on the thermal resistance from the component to the environment, and the heat dissipated by the component. To ensure that the component temperature does not overheat, a thermal engineer seeks to find an efficient heat transfer path from the device to the environment. The heat transfer path may be from the component to a printed circuit board (PCB), to a heat sink, to air flow provided by a fan, but in all instances, eventually to the environment.

Cooling system of an Asus GTX-650 graphics card; three heat pipes are visible

Two additional design factors also influence the thermal/mechanical performance of the thermal design:

1. The method by which the heat sink is mounted on a component or processor. This will be discussed under the section *attachment methods.*

2. For each interface between two objects in contact with each other, there will be a temperature drop across the interface. For such composite systems, the temperature drop across the interface may be appreciable. This temperature change may be attributed to what is known as the thermal contact resistance. *Thermal interface materials* (TIM) decrease the thermal contact resistance.

Attachment Methods

As power dissipation of components increases and component package size decreases, thermal engineers must innovate to ensure components won't overheat. Devices that run cooler last longer. A heat sink design must fulfill both its thermal as well as its mechanical requirements. Concerning the latter, the component must remain in thermal contact with its heat sink with reasonable shock and vibration. The heat sink could be the copper foil of a circuit board, or else a separate heat sink mounted onto the component or circuit board. Attachment methods include thermally conductive tape or epoxy, wire-form z clips, flat spring clips, standoff spacers, and push pins with ends that expand after installing.

Thermally Conductive Tape

Thermally conductive tape is one of the most cost-effective heat sink attachment materials. It is suitable for low-mass heat sinks and for components with low power dissipation. It consists of a thermally conductive carrier material with a pressure-sensitive adhesive on each side.

This tape is applied to the base of the heat sink, which is then attached to the component. Following are factors that influence the performance of thermal tape:

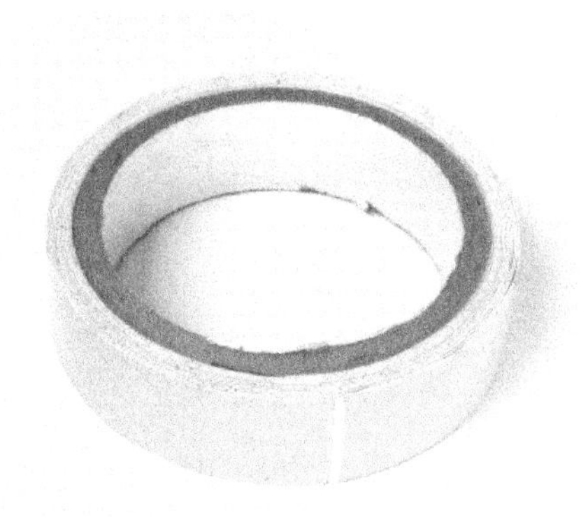

Roll of thermally conductive tape.

1. Surfaces of both the component and heat sink must be clean, with no residue such as a film of silicone grease.

2. Preload pressure is essential to ensure good contact. Insufficient pressure results in areas of non-contact with trapped air, and results in higher-than-expected interface thermal resistance.

3. Thicker tapes tend to provide better "wettability" with uneven component surfaces. "Wettability" is the percentage area of contact of a tape on a component. Thicker tapes, however, have a higher thermal resistance than thinner tapes. From a design standpoint, it is best to strike a balance by selecting a tape thickness that provides maximum "wettablilty" with minimum thermal resistance.

Epoxy

Epoxy is more expensive than tape, but provides a greater mechanical bond between the heat sink and component, as well as improved thermal conductivity. The epoxy chosen must be formulated for this purpose. Most epoxies are two-part liquid formulations that must be thoroughly mixed before being applied to the heat sink, and before the heat sink is placed on the component. The epoxy is then cured for a specified time, which can vary from 2 hours to 48 hours. Faster cure time can be achieved at higher temperatures. The surfaces to which the epoxy is applied must be clean and free of any residue.

The epoxy bond between the heat sink and component is semi-permanent/permanent. This makes re-work very difficult and at times impossible. The most typical damage caused by rework is the separation of the component die heat spreader from its package.

More expensive than tape and epoxy, wire form z-clips attach heat sinks mechanically. To use the z-clips, the printed circuit board must have anchors. Anchors can be either soldered onto the board, or pushed through. Either type requires holes to be designed into the board. The use of RoHS solder must be allowed for because such solder is mechanically weaker than traditional Pb/Sn solder.

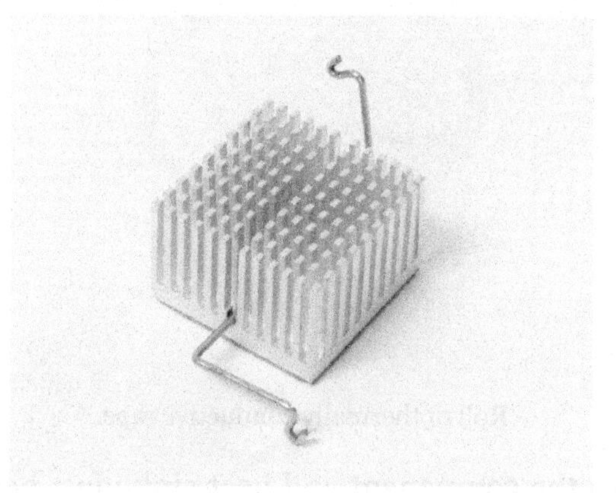

A pin fin heat sink with a Z-clip retainer.

Wire form Z-clips

To assemble with a z-clip, attach one side of it to one of the anchors. Deflect the spring until the other side of the clip can be placed in the other anchor. The deflection develops a spring load on the component, which maintains very good contact. In addition to the mechanical attachment that the z-clip provides, it also permits using higher-performance thermal interface materials, such as phase change types.

Two heat sink attachment methods, namely the maxiGRIP (left) and Talon Clip (right).

Clips

Available for processors and ball grid array (BGA) components, clips allow the attachment of a BGA heat sink directly to the component. The clips make use of the gap created by the ball grid array (BGA) between the component underside and PCB top surface. The clips therefore require no holes in the PCB. They also allow for easy rework of components. Examples of commercially available clips are the maxiGRIP™ and superGRIP™ range from Advanced Thermal Solutions (ATS) and the Talon Clip™ from Malico. The three aforementioned clipping methods use plastic frames for the clips, but the ATS designs uses metal spring clips to provide the compression force. The Malico design uses the plastic “arm” to provide a mechanical load on the component. Depending on the product requirement, the clipping methods will have to meet shock and vibration standards, such as Telecordia GR-63-CORE, ETSI 300 019 and MIL-STD-810.

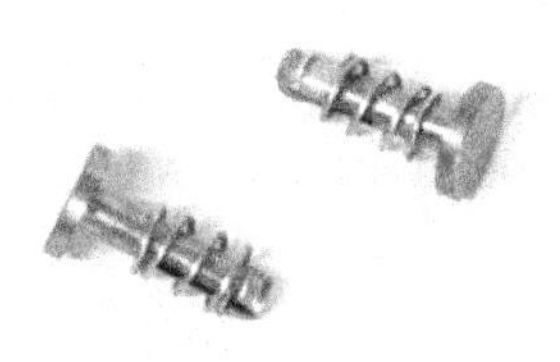

A pair of push pins.

Push Pins with Compression Springs

For larger heat sinks and higher preloads, push pins with compression springs are very effective. The push pins, typically made of brass or plastic, have a flexible barb at the end that engages with a hole in the PCB; once installed, the barb retains the pin. The compression spring holds the assembly together and maintains contact between the heat sink and component. Care is needed in selection of push pin size. Too great an insertion force can result in the die cracking and consequent component failure.

Threaded Standoffs with Compression Springs

For very large heat sinks, there is no substitute for the threaded standoff and compression spring attachment method. A threaded standoff is essentially a hollow metal tube with internal threads. One end is secured with a screw through a hole in the PCB. The other end accepts a screw which compresses the spring, completing the assembly. A typical heat sink assembly uses two to four standoffs, which tends to make this the most costly heat sink attachment design. Another disadvantage is the need for holes in the PCB.

Summary of heat sink attachment methods

Method	Pros	Cons	Cost
Thermal tape	Easy to attach. Inexpensive.	Cannot provide mechanical attachment for heavier heat sinks or for high vibration environments. Surface must be cleaned for optimal adhesion. Moderate to low thermal conductivity.	Very low
Epoxy	Strong mechanical adhesion. Relatively inexpensive.	Makes board rework difficult since it can damage component. Surface must be cleaned for optimal adhesion.	Very low

Wire form Z-clips	Strong mechanical attachment. Easy removal/rework. Applies a preload to the thermal interface material, improving thermal performance.	Requires holes in the board or solder anchors. More expensive than tape or epoxy. Custom designs.	Low
Clip-on	Applies a preload to the thermal interface material, improving thermal performance. Requires no holes or anchors. Easy removal/ rework.	Must have "keep out" zone around the BGA for the clip. Extra assembly steps.	Low
Push pin with compression springs	Strong mechanical attachment. Highest thermal interface material preload. Easy removal and installation.	Requires holes in the board which increases complexity of traces in PCB.	Moderate
Stand-offs with compression springs	Strongest mechanical attachment. Highest preload for the thermal interface material. Ideal for large heat sinks.	Requires holes in the board which increases complexity of trace layout. Complicated assembly.	High

Thermal Interface Materials

Thermal conductivity and the interface resistance form part of the thermal interface resistance of a thermal interface material.

Thermal contact resistance occurs due to the voids created by surface roughness effects, defects and misalignment of the interface. The voids present in the interface are filled with air. Heat transfer is therefore due to conduction across the actual contact area and to conduction (or natural convection) and radiation across the gaps. If the contact area is small, as it is for rough surfaces, the major contribution to the resistance is made by the gaps. To decrease the thermal contact resistance, the surface roughness can be decreased while the interface pressure is increased. However, these improving methods are not always practical or possible for electronic equipment. Thermal interface materials (TIM) are a common way to overcome these limitations,

Properly applied thermal interface materials displace the air that is present in the gaps between the two objects with a material that has a much-higher thermal conductivity. Air has a thermal conductivity of 0.022 W/m•K while TIMs have conductivities of 0.3 W/m•K and higher.

When selecting a TIM, care must be taken with the values supplied by the manufacturer. Most manufacturers give a value for the thermal conductivity of a material. However, the thermal conductivity does not take into account the interface resistances. Therefore, if a TIM has a high thermal conductivity, it does not necessarily mean that the interface resistance will be low.

Selection of a TIM is based on three parameters: the interface gap which the TIM must fill, the contact pressure, and the electrical resistivity of the TIM. The contact pressure is the pressure applied to the interface between the two materials. The selection does not include the cost of the material. Electrical resistivity may be important depending upon electrical design details.

Selection Based on Interface Gap

Interface gap values		**Products types available**
< 0.05 mm	< 2 mil	Thermal grease, epoxy, phase change materials
0.05 – 0.1 mm	2 – 5 mil	Phase change materials, polyimide, graphite or aluminium tapes
0.1 - 0,5 mm	5 – 18 mil	Silicone-coated fabrics
> 0.5 mm	> 18 mil	Gap fillers

Selection Based on Contact Pressure

Contact pressure scale	**Typical pressure ranges**	**Product types available**
Very low	< 70 kPa	Gap fillers
Low	< 140 kPa	Thermal grease, epoxy, polyimide, graphite or aluminium tapes
High	2 MPa	Silicone-coated fabrics

Selection Based on Dielectric Strength

Electrical insulation	**Dielectric strength**	**Typical values**		**Product types available**
Not required	N/A	N/A	N/A	Thermal grease, epoxy, phase change materials, graphite or aluminium tapes.
Required	Low	10 kV/mm	< 300 V/mil	Silicone coated fabrics, gap fillers
Required	High	60 kV/mm	> 1500 V/mil	Polyimide tape

TIM Application Notes Based on Product Type

Product type	**Application notes**	**Thermal performance**
Thermal paste	Messy. Labor-intensive. Relatively long assembly time.	++++
Epoxy	Creates "permanent" interface bond.	++++

Phase change	Allows for pre-attachment. Softens and conforms to interface defects at operational temperatures. Can be repositioned in field.	++++
Thermal tapes, including graphite, polyimide, and aluminium tapes	Easy to apply. Some mechanical strength.	+++
Silicone coated fabrics	Provide cushioning and sealing while still allowing heat transfer.	+
Gap filler	Can be used to thermally couple differing-height components to a heat spreader or heat sink. Naturally tacky.	++

High power LEDs from Philips Lumileds Lighting Company mounted on 21 mm star shaped aluminium-core PCBs

Light-emitting Diode Lamps

Light-emitting diode (LED) performance and lifetime are strong functions of their temperature. Effective cooling is therefore essential. A case study of a LED based downlighter shows an example of the calculations done in order to calculate the required heat sink necessary for the effective cooling of lighting system. The article also shows that in order to get confidence in the results, multiple independent solutions are required that give similar results. Specifically, results of the experimental, numerical and theoretical methods should all be within 10% of each other to give high confidence in the results.

In Soldering

Temporary heat sinks are sometimes used while soldering circuit boards, preventing excessive heat from damaging sensitive nearby electronics. In the simplest case, this means partially gripping a component using a heavy metal crocodile clip, hemostat or similar clamp. Modern semiconductor devices, which are designed to be assembled by reflow soldering, can usually tolerate soldering temperatures without damage. On the other hand, electrical components such as magnetic reed switches can malfunction if exposed to hotter soldering irons, so this practice is still very much in use.

Methods to Determine Performance

In general, a heat sink performance is a function of material thermal conductivity, dimensions, fin type, heat transfer coefficient, air flow rate, and duct size. To determine the thermal performance of a heat sink, a theoretical model can be made. Alternatively, the thermal performance can be measured experimentally. Due to the complex nature of the highly 3D flow in present applications, numerical methods or computational fluid dynamics (CFD) can also be used. This section will discuss the aforementioned methods for the determination of the heat sink thermal performance.

A Heat Transfer Theoretical Model

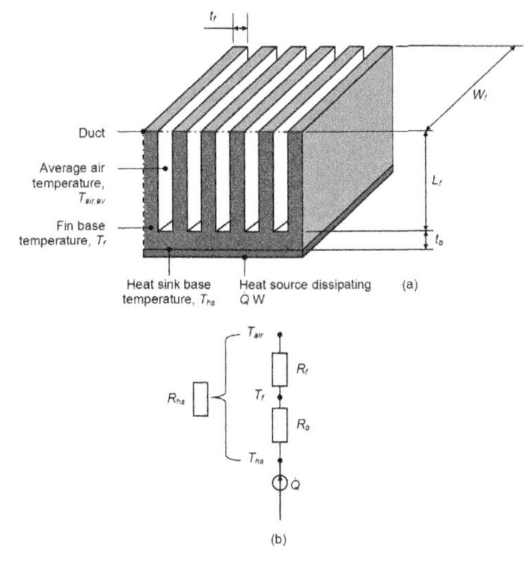

Sketch of a heat sink with equivalent thermal resistances.

Thermal resistance and heat transfer coefficient plotted against flow rate for the specific heat sink design used in. The data was generated using the equations provided in the article. The data shows that for an increasing air flow rate, the thermal resistance of the heat sink decreases.

One of the methods to determine the performance of a heat sink is to use heat transfer and fluid dynamics theory. One such method has been published by Jeggels, et al., though this work is limited to ducted flow. Ducted flow is where the air is forced to flow through a channel which fits tightly over the heat sink. This makes sure that all the air goes through the channels formed by the fins of the heat sink. When the air flow is not ducted, a certain percentage of air flow will bypass the heat sink. Flow bypass was found to increase with increasing fin density and clearance, while remaining relatively insensitive to inlet duct velocity.

The heat sink thermal resistance model consists of two resistances, namely the resistance in the heat sink base, R_b, and the resistance in the fins, R_f. The heat sink base thermal resistance, R_b, can be written as follows if the source is a uniformly applied the heat sink base. If it is not, then the base resistance is primarily spreading resistance:

$$R_b = \frac{t_b}{kA_b} \quad (4)$$

where k is the heat sink base thickness, A_b is the heat sink material thermal conductivity and R_f is the area of the heat sink base.

The thermal resistance from the base of the fins to the air, , can be calculated by the following formulas.

$$R_f = \frac{1}{nh_f W_f \left(t_f + 2\eta_f L_f\right)} \quad (5)$$

$$\eta_f = \frac{\tanh mL_c}{mL_c} \quad (6)$$

$$mL_c = \sqrt{\frac{2h_f}{kt_f}} L_f \quad (7)$$

$$D_h = \frac{4A_{ch}}{P_{ch}} \quad (8)$$

$$Re = \frac{4\dot{G}\rho}{n\pi D_h \mu} \quad (9)$$

$$f = (0.79 \ln Re - 1.64)^{-2} \quad (10)$$

$$Nu = \frac{(f/8)(Re-1000)Pr}{1+12.7(f/8)^{0.5}(Pr^{\frac{2}{3}}-1)} \quad (11)$$

$$h_f = \frac{Nuk_{air}}{D_h} \quad (12)$$

$$\rho = \frac{P_{atm}}{R_a T_{in}} \quad (13)$$

The flow rate can be determined by the intersection of the heat sink system curve and the fan curve. The heat sink system curve can be calculated by the flow resistance of the channels and inlet and outlet losses as done in standard fluid mechanics text books, such as Potter, et al. and White.

Once the heat sink base and fin resistances are known, then the heat sink thermal resistance, R_{hs} can be calculated as: $R_{hs} = R_b + R_f$ (14).

Using the equations 5 to 13 and the dimensional data in, the thermal resistance for the fins was calculated for various air flow rates. The data for the thermal resistance and heat transfer coefficient are shown in the diagram, which shows that for an increasing air flow rate, the thermal resistance of the heat sink increases.

Experimental Methods

Experimental tests are one of the more popular ways to determine the heat sink thermal performance. In order to determine the heat sink thermal resistance, the flow rate, input power, inlet air temperature and heat sink base temperature need to be known. Vendor-supplied data is commonly provided for ducted test results. However, the results are optimistic and can give misleading data when heat sinks are used in an unducted application. More details on heat sink testing methods and common oversights can be found in Azar, et al.

Numerical Methods

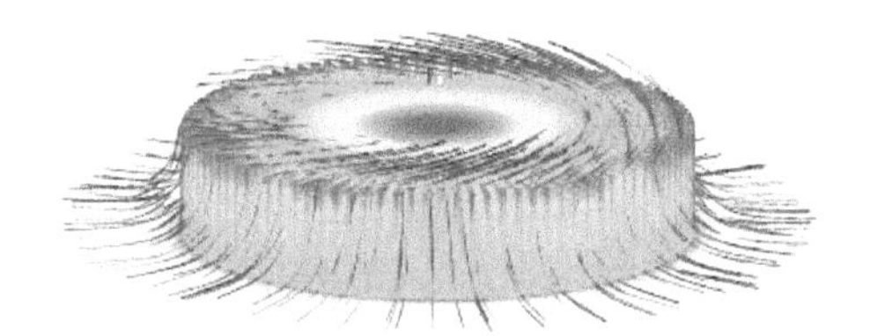

Radial heat sink with thermal profile and swirling forced convection flow trajectories predicted using a CFD analysis package

In industry, thermal analyses are often ignored in the design process or performed too late — when design changes are limited and become too costly. Of the three methods mentioned in this article, theoretical and numerical methods can be used to determine an estimate of the heat sink or component temperatures of products before a physical model has been made. A theoretical model is normally used as a first order estimate. Online heat sink calculators can provide a reasonable estimate of forced and natural convection heat sink performance based on a combination of theoretical and empirically derived correlations. Numerical methods or computational fluid dynamics (CFD) provide a qualitative (and sometimes even quantitative) prediction of fluid flows. What this means is that it will give a visual or post-processed result of a simulation, like the images in figures 16 and 17, and the CFD animations in figure 18 and 19, but the quantitative or absolute accuracy of the result is sensitive to the inclusion and accuracy of the appropriate parameters.

CFD can give an insight into flow patterns that are difficult, expensive or impossible to study using experimental methods. Experiments can give a quantitative description of flow phenomena using measurements for one quantity at a time, at a limited number of points and time instances. If a full-scale model is not available or not practical, scale models or dummy models can be used. The experiments can have a limited range of problems and operating conditions. Simulations can give a prediction of flow phenomena using CFD software for all desired quantities, with high resolution in space and time and virtually any problem and realistic operating conditions. However, if critical, the results may need to be validated.

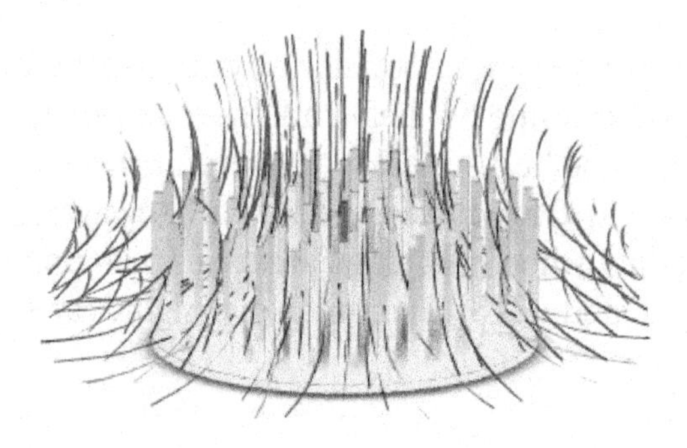

Pin fin heat sink with thermal profile and free convection flow trajectories predicted using a CFD analysis package

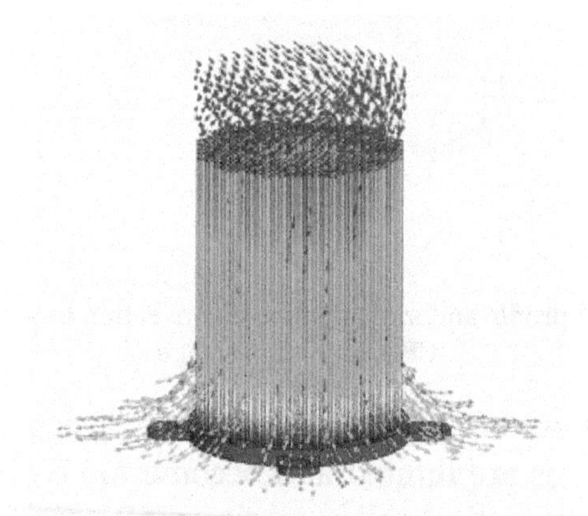

38mm diameter by 50mm tall pin fin heat sink with thermal profile and swirling animated forced convection flow trajectories from a vaneaxial fan, predicted using a CFD analysis package

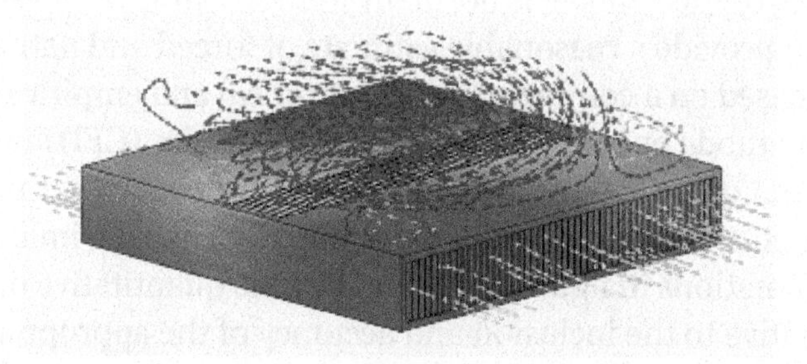

60mm by 60mm by 10mm straight finned heat sink with thermal profile and swirling animated forced convection flow trajectories from a tubeaxial fan, predicted using a CFD analysis package

Heat Pipe

A laptop heat pipe system

A heat pipe is a heat-transfer device that combines the principles of both thermal conductivity and phase transition to efficiently manage the transfer of heat between two solid interfaces.

At the hot interface of a heat pipe a liquid in contact with a thermally conductive solid surface turns into a vapor by absorbing heat from that surface. The vapor then travels along the heat pipe to the cold interface and condenses back into a liquid – releasing the latent heat. The liquid then returns to the hot interface through either capillary action, centrifugal force, or gravity, and the cycle repeats. Due to the very high heat transfer coefficients for boiling and condensation, heat pipes are highly effective thermal conductors. The effective thermal conductivity varies with heat pipe length, and can approach 100 kW/(m·K) for long heat pipes, in comparison with approximately 0.4 kW/(m·K) for copper.

Structure, Design and Construction

A typical heat pipe consists of a sealed pipe or tube made of a material that is compatible with the working fluid such as copper for water heat pipes, or aluminium for ammonia heat pipes. Typically, a vacuum pump is used to remove the air from the empty heat pipe. The heat pipe is partially filled with a *working fluid* and then sealed. The working fluid mass is chosen so that the heat pipe contains both vapor and liquid over the operating temperature range.

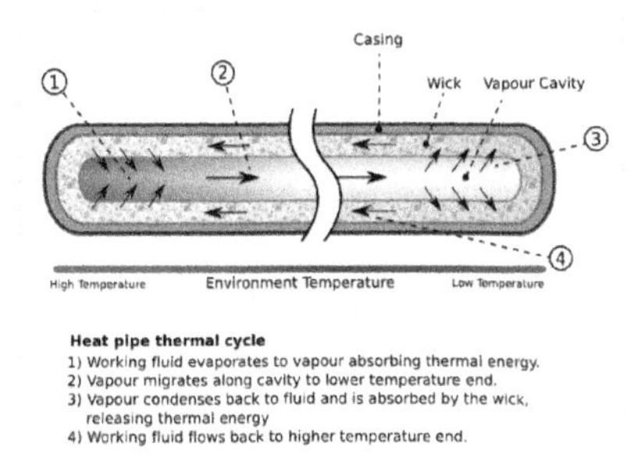

Diagram showing components and mechanism for a heat pipe containing a wick

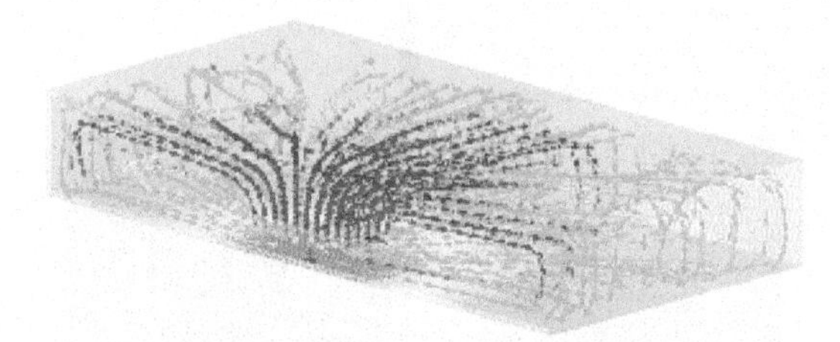

This 100 mm by 100 mm by 10 mm high thin flat heat pipe (heat spreader) animation was created using high resolution CFD analysis and shows temperature contoured flow trajectories, predicted using a CFD analysis package, courtesy of NCI.

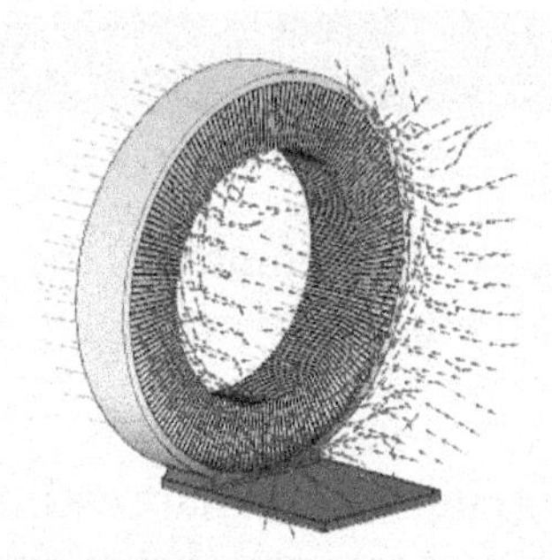

This 120 mm diameter vapor chamber (heat spreader) heat sink design thermal animation was created using high resolution CFD analysis and shows temperature contoured heat sink surface and fluid flow trajectories predicted using a CFD analysis package, courtesy of NCI.

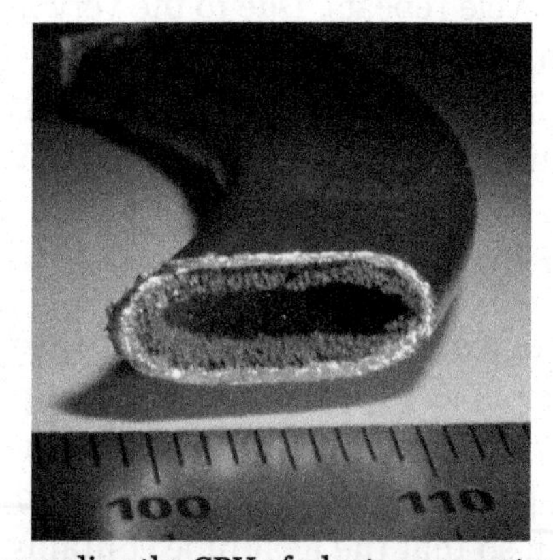

Cross section of a heat pipe for cooling the CPU of a laptop computer. Ruler scale is in millimeters.

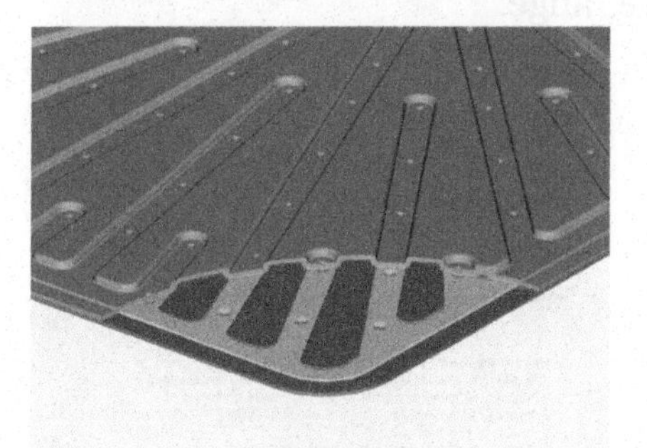

Cut-away view of a 500 μm thick flat heat pipe with a thin planar capillary (aqua colored)

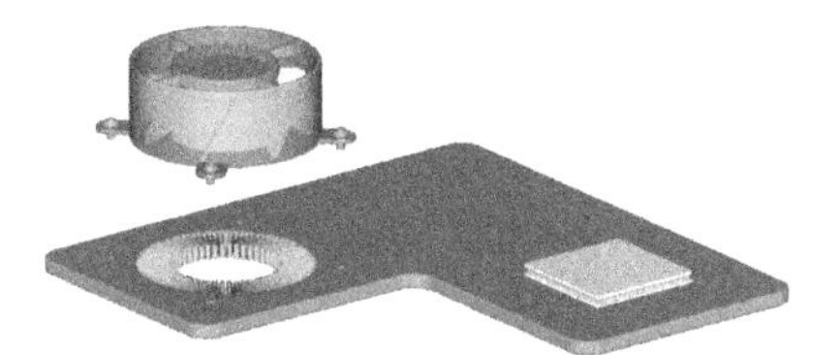

Thin flat heat pipe (heat spreader) with remote heat sink and fan

Below the operating temperature, the liquid is too cold and cannot vaporize into a gas. Above the operating temperature, all the liquid has turned to gas, and the environmental temperature is too high for any of the gas to condense. Whether too high or too low, thermal conduction is still possible through the walls of the heat pipe, but at a greatly reduced rate of thermal transfer.

Working fluids are chosen according to the temperatures at which the heat pipe must operate, with examples ranging from liquid helium for extremely low temperature applications (2–4 K) to mercury (523–923 K), sodium (873–1473 K) and even indium (2000–3000 K) for extremely high temperatures. The vast majority of heat pipes for room temperature applications use ammonia (213–373 K), alcohol (methanol (283–403 K) or ethanol (273–403 K)) or water (298–573 K) as the working fluid. Copper/water heat pipes have a copper envelope, use water as the working fluid and typically operate in the temperature range of 20 to 150 °C. Water heat pipes are sometimes filled by partially filling with water, heating until the water boils and displaces the air, and then sealed while hot.

For the heat pipe to transfer heat, it must contain saturated liquid and its vapor (gas phase). The saturated liquid vaporizes and travels to the condenser, where it is cooled and turned back to a saturated liquid. In a standard heat pipe, the condensed liquid is returned to the evaporator using a wick structure exerting a capillary action on the liquid phase of the working fluid. Wick structures used in heat pipes include sintered metal powder, screen, and grooved wicks, which have a series of grooves parallel to the pipe axis. When the condenser is located above the evaporator in a gravitational field, gravity can return the liquid. In this case, the heat pipe is a thermosyphon. Finally, rotating heat pipes use centrifugal forces to return liquid from the condenser to the evaporator.

Heat pipes contain no mechanical moving parts and typically require no maintenance, though non-condensable gases that diffuse through the pipe's walls, resulting from breakdown of the working fluid or as impurities extant in the material, may eventually reduce the pipe's effectiveness at transferring heat.

The advantage of heat pipes over many other heat-dissipation mechanisms is their great efficiency in transferring heat. A pipe one inch in diameter and two feet long can transfer 12,500 BTU (3.7 kWh) per hour at 1,800 °F (980 °C) with only 18 °F (10 °C) drop from end to end. Some heat pipes have demonstrated a heat flux of more than 23 kW/cm^2, about four times the heat flux through the surface of the sun.

Heat pipe materials and working fluids

Heat pipes have an envelope, a wick, and a working fluid. Heat pipes are designed for very long term operation with no maintenance, so the heat pipe wall and wick must be compatible with the working fluid. Some material/working fluids pairs that appear to be compatible are not. For example, water in an aluminum envelope will develop large amounts of non-condensable gas over a few hours or days, preventing normal operation of the heat pipe.

Since heat pipes were rediscovered by George Grover in 1963, extensive life tests have been conducted to determine compatible envelope/pairs, some going on for decades. In a heat pipe life test, heat pipes are operated for long periods of time, and monitored for problems such as non-condensable gas generation, material transport, and corrosion.

The most commonly used envelope (and wick)/fluid pairs include:

- Copper envelope/Water working fluid for electronics cooling. This is by far the most common type of heat pipe.
- Copper or Steel envelope/Refrigerant R134a working fluid for energy recovery in HVAC systems
- Aluminum envelope/Ammonia working fluid for Spacecraft Thermal Control
- Superalloy envelope/Alkali Metal (Cesium, Potassium, Sodium) working fluid for high temperature heat pipes, most commonly used for calibrating primary temperature measurement devices

Other pairs include stainless steel envelopes with nitrogen, oxygen, neon, hydrogen, or helium working fluids at temperatures below 100 K, copper/methanol heat pipes for electronics cooling when the heat pipe must operate below the water range, aluminum/ethane heat pipes for spacecraft thermal control in environments when ammonia can freeze, and refractory metal envelope/lithium working fluid for high temperature (above 1050 °C) applications.

Different Types of Heat Pipes

In addition to standard, Constant Conductance Heat Pipes (CCHPs), there are a number of other types of heat pipes, including:

- Vapor Chambers (flat heat pipes), which are used for heat flux transformation, and isothermalization of surfaces
- Variable Conductance Heat Pipes (VCHPs), which use a Non-Condensable Gas (NCG) to change the heat pipe effective thermal conductivity as power or the heat sink conditions change

- Pressure Controlled Heat Pipes (PCHPs), which are a VCHP where the volume of the reservoir, or the NCG mass can be changed, to give more precise temperature control
- Diode Heat Pipes, which have a high thermal conductivity in the forward direction, and a low thermal conductivity in the reverse direction
- Thermosyphons, which are heat pipes where the liquid is returned to the evaporator by gravitational/accelerational forces,
- Rotating heat pipes, where the liquid is returned to the evaporator by centrifugal forces

Vapor Chamber or Flat Heat Pipes

Thin planar heat pipes (heat spreaders) have the same primary components as tubular heat pipes: a hermetically sealed hollow vessel, a working fluid, and a closed-loop capillary recirculation system. In addition, a series of posts are generally used in a vapor chamber, to prevent collapse of the flat top and bottom when the pressure is lower than atmospheric, which is 100 °C for water vapor chambers.

There are two main applications for vapor chambers. First, they are used when high powers and heat fluxes are applied to a relatively small evaporator. Heat input to the evaporator vaporizes liquid, which flows in two dimensions to the condenser surfaces. After the vapor condenses on the condenser surfaces, capillary forces in the wick return the condensate to the evaporator. Note that most vapor chambers are insensitive to gravity, and will still operate when inverted, with the evaporator above the condenser. In this application, the vapor chamber acts as a heat flux transformer, cooling a high heat flux from an electronic chip or laser diode, and transforming it to a lower heat flux that can be removed by natural or forced convection. With special evaporator wicks, vapor chambers can remove 2000 W over 4 cm^2, or 700 W over 1 cm^2.

Second, compared to a one-dimensional tubular heat pipe, the width of a two-dimensional heat pipe allows an adequate cross section for heat flow even with a very thin device. These thin planar heat pipes are finding their way into "height sensitive" applications, such as notebook computers and surface mount circuit board cores. These vapor chambers are typically fabricated from aluminum extrusions, and use acetone as the working fluid. It is possible to produce flat heat pipes as thin as 1.0 mm (slightly thicker than a 0.76 mm credit card).

Variable Conductance Heat Pipes (VCHPs)

Standard heat pipes are constant conductance devices, where the heat pipe operating temperature is set by the source and sink temperatures, the thermal resistances from the source to the heat pipe, and the thermal resistances from the heat pipe to the sink. In these heat pipes, the temperature drops linearly as the power or condenser temperature is reduced. For some applications, such as satellite or research balloon thermal control, the electronics will be overcooled at low powers, or at the low sink temperatures. Variable Conductance Heat Pipes (VCHPs) are used to passively maintain the temperature of the electronics being cooled as power and sink conditions change.

VCHPs have two additions compared to a standard heat pipe: 1. A reservoir, and 2. A Non-Condensable Gas (NCG) added to the heat pipe, in addition to the working fluid. This NCG is typically argon for standard VCHPs, and helium for thermosyphons. When the heat pipe is not operating, the NCG and working fluid vapor are mixed throughout the heat pipe vapor space. When the VCHP is operating, the NCG is swept toward the condenser end of the heat pipe by the flow of the working fluid vapor. Most of the NCG is located in the reservoir, while the remainder blocks a portion of the heat pipe condenser. The VCHP works by varying the active length of the condenser. When the power or heat sink temperature is increased, the heat pipe vapor temperature and pressure increase. The increased vapor pressure forces more of the NCG into the reservoir, increasing the active condenser length and the heat pipe conductance. Conversely, when the power or heat sink temperature is decreased, the heat pipe vapor temperature and pressure decrease, and the NCG expands, reducing the active condenser length and heat pipe conductance. The addition of a small heater on the reservoir, with the power controlled by the evaporator temperature, will allow thermal control of roughly ±1-2 °C. In one example, the evaporator temperature was maintained in a ±1.65 °C control band, as power was varied from 72 to 150 W, and heat sink temperature varied from +15 °C to -65 °C.

Pressure Controlled Heat Pipes (PCHPs) can be used when tighter temperature control is required. In a PCHP, the evaporator temperature is used to either vary the reservoir volume, or the amount of NCG in the heat pipe. PCHPs have shown milli-Kelvin temperature control.

Diode Heat Pipes

Conventional heat pipes transfer heat in either direction, from the hotter to the colder end of the heat pipe. Several different heat pipes act as a thermal diode, transferring heat in one direction, while acting as an insulator in the other:

- Thermosyphons, which only transfer heat from the bottom to the top of the thermosyphon, where the condensate returns by gravity. When the thermosyphon is heated at the top, there is no liquid available to evaporate.
- Rotating Heat Pipes, where the heat pipe is shaped so that liquid can only travel by centrifugal forces from the nominal evaporator to the nominal condenser. Again, no liquid is available when the nominal condenser is heated.
- Vapor Trap Diode Heat Pipes.

- Liquid Trap Diode Heat Pipes.

A Vapor Trap Diode is fabricated in a similar fashion to a Variable Conductance Heat Pipe (VCHP), with a gas reservoir at the end of the condenser. During fabrication, the heat pipe is charged with the working fluid and a controlled amount of a Non-Condensable Gas (NCG). During normal operation, the flow of the working fluid vapor from the evaporator to the condenser sweeps the NCG into the reservoir, where it doesn't interfere with the normal heat pipe operation. When the nominal condenser is heated, the vapor flow is from the nominal condenser to the nominal evaporator. The NCG is dragged along with the flowing vapor, completely blocking the nominal evaporator, and greatly increasing the thermal resistivity of the heat pipe. In general, there is some heat transfer to the nominal adiabatic section. Heat is then conducted through the heat pipe walls to the evaporator. In one example, a vapor trap diode carried 95 W in the forward direction, and only 4.3 W in the reverse direction.

A Liquid Trap Diode has a wicked reservoir at the evaporator end of the heat pipe, with a separate wick that is not in communication with the wick in the remainder of the heat pipe. During normal operation, the evaporator and reservoir are heated. The vapor flows to the condenser, and liquid returns to the evaporator by capillary forces in the wick. The reservoir eventually dries out, since there is no method for returning liquid. When the nominal condenser is heated, liquid condenses in the evaporator and the reservoir. While the liquid can return to the nominal condenser from the nominal evaporator, the liquid in the reservoir is trapped, since the reservoir wick is not connected. Eventually, all of the liquid is trapped in the reservoir, and the heat pipe ceases operation.

Thermosyphons

Most heat pipes use a wick and capillary action to return the liquid from the condenser to the evaporator. The liquid is sucked up to the evaporator, similar to the way that a sponge sucks up water when an edge is placed in contact with a water pool. The wick allows the heat pipe to operate in any orientation, but the maximum adverse elevation (evaporator over condenser) is relatively small, on the order of 25 cm long for a typical water heat pipe.

Taller heat pipes must be gravity aided. When the evaporator is located below the condenser, the liquid can drain back by gravity instead of requiring a wick. Such a grav-ity aided heat pipe is known as a thermosyphon. Please note that a heat pipe thermosyphon is different than a thermosiphon, which transfers heat by single phase natural convection heat transfer in a loop.

In a thermosyphon, liquid working fluid is vaporized by a heat supplied to the evaporator at the bottom of the heat pipe. The vapor travels to the condenser at the top of the heat pipe, where it condenses. The liquid then drains back to the bottom of the heat pipe by gravity, and the cycle repeats. Thermosyphons also act as diode heat pipes. When heat is applied to the condenser, there is no condensate available, and hence no way to form vapor and transfer heat to the evaporator.

While a typical terrestrial water heat pipe is less than 30 cm long, thermosyphons are often several meters long. As discussed below, the thermosyphons used to cool the Alaska pipe line were roughly 11 to 12 m long. Even longer thermosyphons have been proposed for the extraction of geothermal energy. For example, Storch et al. fabricated a 53 mm I.D., 92 m long propane thermosyphon that carried roughly 6 kW of heat.

Loop Heat Pipe

A loop heat pipe (LHP) is a passive two-phase transfer device related to the heat pipe. It can carry higher power over longer distances by having co-current liquid and vapor flow, in contrast to the counter-current flow in a heat pipe. This allows the wick in a loop heat pipe to be required only in the evaporator and compensation chamber. Micro loop heat pipes have been developed and successfully employed in a wide sphere of applications both on the ground and in space.

Heat Transfer

A heat sink (aluminium) with heat pipe (copper)

Heat pipes employ evaporative cooling to transfer thermal energy from one point to another by the evaporation and condensation of a working fluid or coolant. Heat pipes rely on a temperature difference between the ends of the pipe, and cannot lower temperatures at either end beyond the ambient temperature (hence they tend to equalise the temperature within the pipe).

When one end of the heat pipe is heated the working fluid inside the pipe at that end evaporates and increases the vapour pressure inside the cavity of the heat pipe. The latent heat of evaporation absorbed by the vaporisation of the working fluid reduces the temperature at the hot end of the pipe.

The vapour pressure over the hot liquid working fluid at the hot end of the pipe is higher than the equilibrium vapour pressure over the condensing working fluid at the cooler end of the pipe, and this pressure difference drives a rapid mass transfer to the condensing end where the excess vapour condenses, releases its latent heat, and warms the cool end of the pipe. Non-condensing gases (caused by contamination for instance) in the vapour impede the gas flow and reduce the effectiveness of the heat pipe, particularly at low temperatures, where vapour pressures are low. The speed of molecules in a gas is approximately the speed of sound, and in the absence of noncondensing gases (i.e., if there is only a gas phase present) this is the upper limit to the velocity with which they could travel in the heat pipe. In practice, the speed of the vapour through the heat pipe is limited by the rate of condensation at the cold end and far lower than the molecular speed.

The condensed working fluid then flows back to the hot end of the pipe. In the case of vertically oriented heat pipes the fluid may be moved by the force of gravity. In the case of heat pipes containing wicks, the fluid is returned by capillary action.

When making heat pipes, there is no need to create a vacuum in the pipe. One simply boils the working fluid in the heat pipe until the resulting vapour has purged the non-condensing gases from the pipe, and then seals the end.

An interesting property of heat pipes is the temperature range over which they are effective. Initially, it might be suspected that a water-charged heat pipe only works when the hot end reaches the boiling point (100 °C, 212 °F) and steam is transferred to the cold end. However, the boiling point of water depends on the absolute pressure inside the pipe. In an evacuated pipe, water vaporizes from its melting point (0 °C, 32 °F) to its critical point (374 °C; 705 °F), as long as the heat pipe contains both liquid and vapor. Thus a heat pipe can operate at hot-end temperatures as low as just slightly warmer than the melting point of the working fluid, although the maximum power is low at temperatures below 25 °C (77 °F). Similarly, a heat pipe with water as a working fluid can work well above the boiling point (100 °C, 212 °F). The maximum temperature for long term water heat pipes is 270 °C (518 °F), with heat pipes operating up to 300 °C (572 °F) for short term tests.

The main reason for the effectiveness of heat pipes is the evaporation and condensation of the working fluid. The heat of vaporization greatly exceeds the sensible heat capacity. Using water as an example, the energy needed to evaporate one gram of water is 540 times the amount of energy needed to raise the temperature of that same one gram of water by 1 °C. Almost all of that energy is rapidly transferred to the "cold" end when the fluid condenses there, making a very effective heat transfer system with no moving parts.

Development

The general principle of heat pipes using gravity, commonly classified as two phase thermosiphons, dates back to the steam age and Angier March Perkins and his son Loftus Perkins and the "Perkins Tube", which saw widespread use in locomotive boilers and working ovens. Capillary-based heat pipes were first suggested by R.S. Gaugler of General Motors in 1942, who patented the idea, but did not develop it further.

George Grover independently developed capillary-based heat pipes at Los Alamos National Laboratory in 1963, with his patent of that year being the first to use the term "heat pipe", and he is often referred to as "the inventor of the heat pipe". He noted in his notebook:

Such a closed system, requiring no external pumps, may be of particular interest in space reactors in moving heat from the reactor core to a radiating system. In the absence of gravity, the forces must only be such as to overcome the capillary and the drag of the returning vapor through its channels.

Grover's suggestion was taken up by NASA, which played a large role in heat pipe development in the 1960s, particularly regarding applications and reliability in space flight. This was understandable given the low weight, high heat flux, and zero power draw of heat pipes – and that they would not be adversely affected by operating in a zero gravity environment.

The first application of heat pipes in the space program was the thermal equilibration of satellite transponders.As satellites orbit, one side is exposed to the direct radiation of the sun while the opposite side is completely dark and exposed to the deep cold of outer space. This causes severe discrepancies in the temperature (and thus reliability and accuracy) of the transponders. The heat pipe cooling system designed for this purpose managed the high heat fluxes and demonstrated flawless operation with and without the influence of gravity. The cooling system developed was the first use of variable conductance heat pipes to actively regulate heat flow or evaporator temperature.

Wider Usage

NASA has tested heat pipes designed for extreme conditions, with some using liquid sodium metal as the working fluid. Other forms of heat pipes are currently used to cool communication satellites. Publications in 1967 and 1968 by Feldman, Eastman, and Katzoff first discussed applications of heat pipes for wider uses such as in air conditioning, engine cooling, and electronics cooling. These papers were also the first to mention flexible, arterial, and flat plate heat pipes. Publications in 1969 introduced the concept of the rotational heat pipe with its applications to turbine blade cooling and contained the first discussions of heat pipe applications to cryogenic processes.

Starting in the 1980s Sony began incorporating heat pipes into the cooling schemes for some of its commercial electronic products in place of both forced convection and passive finned heat sinks. Initially they were used in receivers and amplifiers, soon spreading to other high heat flux electronics applications.

During the late 1990s increasingly high heat flux microcomputer CPUs spurred a three-fold increase in the number of U.S. heat pipe patent applications. As heat pipes evolved from a specialized industrial heat transfer component to a consumer commodity most development and production moved from the U.S. to Asia.

Modern CPU heat pipes are typically made of copper and use water as the working fluid.

Applications

Spacecraft

Heat pipes on spacecraft typically use a grooved aluminum extrusion as the envelope

The spacecraft thermal control system has the function to keep all components on the spacecraft within their acceptable temperature range. This is complicated by the following:

- Widely varying external conditions, such as eclipses
- Micro-g environment
- Heat removal from the spacecraft by thermal radiation only
- Limited electrical power available, favoring passive solutions
- Long lifetimes, with no possibility of maintenance

Some spacecraft are designed to last for 20 years, so heat transport without electrical power or moving parts is desirable. Rejecting the heat by thermal radiation means that large radiator panes (multiple square meters) are required. Heat pipes and loop heat pipes are used extensively in spacecraft, since they don't require any power to operate, operate nearly isothermally, and can transport heat over long distances.

Grooved wicks are used in spacecraft heat pipes, as shown in the first photograph on the right. The heat pipes are formed by extruding aluminum, and typically have an integral flange to increase the heat transfer area, which lowers the temperature drop. Grooved wicks are used in spacecraft, instead of the screen or sintered wicks used for terrestrial heat pipes, since the heat pipes don't have to operate against gravity in space. This allows spacecraft heat pipes to be several meters long, in contrast to the roughly 25 cm maximum length for a water heat pipe operating on Earth. Ammonia is the most common working fluid for spacecraft heat pipes. Ethane is used when the heat pipe must operate at temperatures below the ammonia freezing temperature.

The second figure shows a typical grooved aluminum/ammonia Variable Conductance Heat Pipe (VCHP) for spacecraft thermal control. The heat pipe is an aluminum extrusion, similar to that shown in the first figure. The bottom flanged area is the evaporator. Above the evaporator, the flange is machined off to allow the adiabatic section to be bent. The condenser is shown above the adiabatic section. The Non-Condensable Gas (NCG) reservoir is located above the main heat pipe. The valve is removed after filling and sealing the heat pipe. When electric heaters are used on the reservoir, the evaporator temperature can be controlled within ±2 K of the setpoint.

Computer Systems

Heat pipes began to be used in computer systems in the late 1990s, when increased power requirements and subsequent increases in heat emission resulted in greater demands on cooling systems. They are now extensively used in many modern computer systems, typically to move heat away from components such as CPUs and GPUs to heat sinks where thermal energy may be dissipated into the environment.

Solar Thermal

Heat pipes are also widely used in solar thermal water heating applications in combination with evacuated tube solar collector arrays. In these applications, distilled water is commonly used as the heat transfer fluid inside a sealed length of copper tubing that is located within an evacuated glass tube and oriented towards the sun. In connecting pipes, the heat transport occurs in the liquid steam phase because the thermal transfer medium is converted into steam in a large section of the collecting pipeline.

In solar thermal water heating applications, an individual absorber tube of an evacuated tube collector is up to 40% more efficient compared to more traditional "flat plate" solar water collectors. This is largely due to the vacuum that exists within the tube, which slows down convective and conductive heat loss. Relative efficiencies of the evacuated tube system are reduced however, when compared to flat plate collectors because the latter have a larger aperture size and can absorb more solar energy per unit area. This means that while an individual evacuated tube has better insulation (lower conductive and convective losses) due to the vacuum created inside the tube, an array of tubes found in a completed solar assembly absorbs less energy per unit area due to there being less absorber surface area pointed toward the sun because of the rounded design of an evacuated tube collector. Therefore, real world efficiencies of both designs are about the same.

Evacuated tube collectors reduce the need for anti-freeze additives since the vacuum helps slow heat loss. However, under prolonged exposure to freezing temperatures the heat transfer fluid can still freeze and precautions must be taken to ensure that the freezing liquid does not damage the evacuated tube when designing systems for such environments. Properly designed solar thermal water heaters can be frost protected down to more than -3 °C with special additives and are being used in Antarctica to heat water.

Permafrost Cooling

Alaska pipeline support legs cooled by heat pipe thermosyphons to keep permafrost frozen

Building on permafrost is difficult because heat from the structure can thaw the permafrost. Heat pipes are used in some cases to avoid the risk of destabilization. For example, in the Trans-Alaska Pipeline System residual ground heat remaining in the oil as well as heat produced by friction and turbulence in the moving oil could conduct down the pipe's support legs and melt the permafrost on which the supports are anchored. This would cause the pipeline to sink and possibly be damaged. To prevent this, each vertical support member has been mounted with four vertical heat pipe thermosyphons.

The significant feature of a thermosyphon is that it is passive and does not require any external power to operate. During the winter, the air is colder than the ground around the supports. The liquid ammonia at the bottom of the thermosyphon is vaporized by heat absorbed from the ground, cooling the surrounding permafrost and lowering its temperature. During the summer, the thermosyphons stop operating, since there is no liquid ammonia available at the top of the heat pipe, but the extreme cooling during the winter allows the ground to remain frozen.

Heat pipes are also used to keep the permafrost frozen alongside parts of the Qinghai–Tibet Railway where the embankment and track absorb the sun's heat. Vertical heat pipes on either side of relevant formations prevent that heat from spreading any further into the surrounding permafrost.

Depending on application there are several thermosyphon designs: thermoprobe, thermopile, depth thermosyphon, sloped-thermosyphon foundation, flat loop thermosyphon foundation, hybrid flat loop thermosyphon foundation.

Cooking

The first commercial heat pipe product was the "Thermal Magic Cooking Pin" developed by Energy Conversion Systems, Inc. and first sold in 1966. The cooking pins used water as the working fluid. The envelope was stainless steel, with an inner copper layer for compatibility. During operation, one end of the heat pipe is poked through the roast. The other end extends into the oven where it draws heat to the middle of the roast. The high effective conductivity of the heat pipe reduces the cooking time for large pieces of meat by one-half.

The principle has also been applied to camping stoves. The heat pipe transfers a large volume of heat at low temperature to allow goods to be baked and other dishes to be cooked in camping-type situations. An example is the Bakepacker system.

Ventilation Heat Recovery

In heating, ventilation and air-conditioning systems, HVAC, heat pipes are positioned within the supply and exhaust air streams of an air handling system or in the exhaust gases of an industrial process, in order to recover the heat energy.

The device consists of a battery of multi-row finned heat pipe tubes located within both the supply and exhaust air streams. Within the exhaust air side of the heat pipe, the refrigerant evaporates, taking its heat from the extract air. The refrigerant vapour moves towards the cooler end of the tube, within the supply air side of the device, where it condenses and gives up its heat. The condensed refrigerant returns by a combination of gravity and capillary action in the wick. Thus heat is transferred from the exhaust air stream through the tube wall to the refrigerant, and then from the refrigerant through the tube wall to the supply air stream.

Because of the characteristics of the device, better efficiencies are obtained when the unit is positioned upright with the supply air side mounted over the exhaust air side, which allows the liquid refrigerant to flow quickly back to the evaporator aided by the force of gravity. Generally, gross heat transfer efficiencies of up to 75% are claimed by manufacturers.

Nuclear Power Conversion

Grover and his colleagues were working on cooling systems for nuclear power cells for space craft, where extreme thermal conditions are encountered. These alkali metal heat pipes transferred heat from the heat source to a thermionic or thermoelectric converter to generate electricity.

Since the early 1990s, numerous nuclear reactor power systems have been proposed using heat pipes for transporting heat between the reactor core and the power conversion system. The first nuclear reactor to produce electricity using heat pipes was first operated on September 13, 2012 in a demonstration using flattop fission.

Wankel Rotary Combustion Engines

Wankel RCEs do have problems connected to the ignition of mix always taking place in the same part of the housing, this inducing thermal dilatation disparities that reduce power output, impair fuel economy, and accelerate wear. SAE paper 2014-01-2160, by Wei Wu et al., describes: 'A Heat Pipe Assisted Air-Cooled Rotary Wankel Engine for Improved Durability, Power and Efficiency', they obtained a reduction in top engine temperature from 231 °C to 129 °C, and the temperature difference reduced from 159 °C to 18 °C for a typical UAV, small-chamber-displacement, air-cooled engine.

Limitations

Heat pipes must be tuned to particular cooling conditions. The choice of pipe material, size and coolant all have an effect on the optimal temperatures at which heat pipes work.

When heated above a certain temperature, all of the working fluid in the heat pipe vaporizes and the condensation process ceases; in such conditions, the heat pipe's thermal conductivity is effectively reduced to the heat conduction properties of its solid metal casing alone. As most heat pipes are constructed of copper (a metal with high heat conductivity), an overheated heatpipe will generally continue to conduct heat at around 1/80 of the original flux via conduction only rather than conduction and evaporation.

In addition, below a certain temperature, the working fluid will not undergo phase change, and the thermal conductivity is reduced to that of the solid metal casing. One of the key criteria for selecting a working fluid is the desired operational temperature range of the application. The lower temperature limit typically occurs a few degrees above the freezing point of the working fluid.

Most manufacturers cannot make a traditional heat pipe smaller than 3 mm in diameter due to material limitations (though 1.85 mm thin sheets with embedded, flattened heat pipes can be fabricated, as well 1.0 mm thin vapor chambers). Experiments have been conducted with micro heat pipes, which use piping with sharp edges, such as triangular or rhombus-like tubing. In these cases, the sharp edges transfer the fluid through capillary action, and no wick is necessary.

Loop Heat pipe

A loop heat pipe (LHP) is a two-phase heat transfer device that uses capillary action to remove heat from a source and passively move it to a condenser or radiator. LHPs are similar to heat pipes but have the advantage of being able to provide reliable operation over long distance and the ability to operate against gravity. They can transport a large heat load over a long distance with a small temperature difference. Different designs of LHPs ranging from powerful, large size LHPs to miniature LHPs (micro loop heat pipe) have been developed and successfully employed in a wide sphere of applications both ground based as well as space applications.

Construction

The most common coolants used in LHPs are anhydrous ammonia and propylene. LHPs are made by controlling the volumes of the reservoir carefully, condenser and vapor and liquid lines so that liquid is always available to the wick. The reservoir volume and fluid charge are set so that there is always fluid in the reservoir even if the condenser and vapor and liquid lines are completely filled.

Generally small pore size and large capillary pumping capability are necessary in a wick. There must be a balance in the wick pumping capability and the wick permeability when designing a heat pipe or loop heat pipe.

Mechanism

In a loop heat pipe, first the heat enters the evaporator and vaporizes the working fluid at the wick outer surface. The vapor then flows down the system of grooves and then goes to the evaporator and the vapor line towards the condenser, where it condenses as heat is removed by the radiator. The two-phase reservoir (or compensation chamber) at the end of the evaporator is specifically designed to operate at a slightly lower temperature than the evaporator (and the condenser). The lower saturation pressure in the reservoir draws the condensate through the condenser and liquid return line. The fluid then flows into a central pipe where it feeds the wick. A secondary wick hydraulically links the reservoir and the primary wick.

Limitations of Heat Pipes

Heat pipes are excellent heat transfer devices but their sphere of application is mainly confined to transferring relatively small heat loads over relatively short distances when the evaporator and condenser are at same horizontal level. This limitation on the part of the heat pipes is mainly related to the major pressure losses associated with the liquid flow through the porous structure, present along the entire length of the heat pipe and viscous interaction between the vapor and liquid phases, also called entrainment losses. For the applications involving transfer of large heat loads over long distances, the thermal performance of the heat pipes is badly affected by increase in these losses. For the same reason conventional heat pipes are very sensitive to the change in orientation in gravitational field. For the unfavorable slopes in evaporator-above-condenser configuration, the pressure losses due to the mass forces in gravity field adds to the total pressure losses and further affect the efficiency of the heat transfer process.

As a result of these limitations, different solutions involving structural modifications to the conventional heat pipe have been proposed. Some of these modifications incorporate arterial tubes with considerably low hydraulic resistance for liquid return to the heat source (arterial heat pipes), while others provide spatial separation of the vapor and liquid phases of the working fluid at the transportation section (separated line heat pipes).

Though these new forms of heat pipes are able to transfer significant heat flows and can increase heat transport length, they remain very sensitive to spatial orientation relative to gravity. To extend functional possibilities of two-phase systems towards applications involving otherwise inoperable slopes in gravity, the advantages provided by the spatial separation of the transportation line and the usage of non-capillary arteries are combined in the loop scheme. This scheme allows heat pipes to be created with higher heat transfer characteristics while maintaining normal operation in any directional orientation. The loop scheme forms the basis of the physical concept of Two-Phase Loops (TPLs).

Origins

Loop heat pipes were patented in USSR in 1974 by Yury F. Gerasimov and Yury F. Maydanik (Inventor's certificate № 449213), all of the former Soviet Union. The patent for LHPs was filed in the USA in 1982 (Patent № 4515209).

Applications

The first space application occurred aboard a Russian spacecraft in 1989. LHPs are now commonly used in space aboard satellites including; Russian Granat, Obzor spacecraft, Boeing's (Hughes) HS 702 communication satellites, Chinese FY-1C meteorological satellite, NASA's ICESat.

LHPs were first flight demonstrated on the NASA space shuttle in 1997 with STS-83 and STS-94.

Loop heat pipes are important parts of systems for cooling electronic components.

Thermosiphon

Thermosiphon (or thermosyphon) is a method of passive heat exchange, based on natural convection, which circulates a fluid without the necessity of a mechanical pump. Thermosiphoning is used for circulation of liquids and volatile gases in heating and cooling applications such as heat pumps, water heaters, boilers and furnaces. Thermosiphoning also occurs across air temperature gradients such as those utilized in a wood fire chimney or solar chimney.

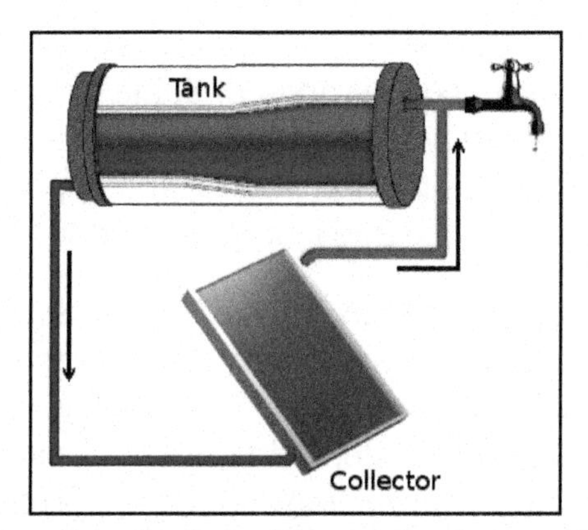

Thermosyphon circulation in a simple solar water heater

This circulation can either be open-loop, as when the substance in a holding tank is passed in one direction via a heated transfer tube mounted at the bottom of the tank to a distribution point—even one mounted above the originating tank—or it can be a vertical closed-loop circuit with return to the original container. Its purpose is to simplify the transfer of liquid or gas while avoiding the cost and complexity of a conventional pump.

Simple Thermosiphon

Natural convection of the liquid starts when heat transfer to the liquid gives rise to a temperature difference from one side of the loop to the other. The phenomenon of thermal expansion means that a temperature difference will have a corresponding difference in density across the loop. The warmer fluid on one side of the loop is less dense and thus more buoyant than the cooler fluid on the other side. The warmer fluid will "float" above the cooler fluid, and the cooler fluid will "sink" below the warmer fluid. This phenomena of natural convection is known by the saying: "heat rises". Convection moves the heated liquid upwards in the system as it is simultaneously replaced by cooler liquid returning by gravity. A good thermosiphon has very little hydraulic resistance so that liquid can flow easily under the relatively low pressure produced by natural convection.

Heat pipes

In some situations the flow of liquid may be reduced further, or stopped, perhaps because the loop is not entirely full of liquid. In this case, the system no longer convects, so it is not a usual "thermosiphon".

Heat can still be transferred in this system by the evaporation and condensation of vapor; however, the system is properly classified as a heat pipe thermosyphon. If the system also contains other fluids, such as air, then the heat flux density will be less than in a real heat pipe, which only contains a single fluid.

The thermosiphon has been sometimes incorrectly described as a 'gravity return heat pipe'. Heat pipes usually have a wick to return the condensate to the evaporator via capillary action. A wick is not needed in a thermosiphon because gravity moves the liquid. The wick allows heat pipes to transfer heat when there is no gravity, which is useful in space. A thermosiphon is "simpler" than a heat pipe.

(Single-phase) thermosiphons can only transfer heat "upward", or away from the acceleration vector. Thus, orientation is much more important for thermosiphons than for heatpipes. Also, thermosiphons can fail because of a bubble in the loop, and require a circulating loop of pipes.

Reboilers and Calandria

If the piping of a thermosiphon resists flow, or excessive heat is applied, the liquid may boil. Since the gas is more buoyant than the liquid, the convective pressure is greater. This is a well known invention called a reboiler. A group of reboilers attached to a pair of plena is called a calandria.

The term "phase change thermosiphon" is a misnomer and should be avoided.When phase change occurs in a thermosiphon, it means that the system either does not have enough fluid, or it is too small to transfer all of the heat by convection alone. To improve the performance, either more fluid is needed (possibly in a larger thermosiphon), or all other fluids (including air) should be pumped out of the loop.

Solar Energy

Solar heating system featuring a thermosiphon

Thermosiphons are used in some liquid-based solar heating systems to heat a liquid such as water. The water is heated passively by solar energy and relies on heat energy being transferred from the sun to a solar collector. The heat from the collector can be transferred to water in two ways: *directly* where water circulates through the collector, or *indirectly* where an anti-freeze solution carries the heat from the collector and transfers it to water in the tank via a heat exchanger. Convection allows for the movement of the heated liquid out of the solar collector to be replaced by colder liquid which is in turn heated. Due to this principle, it is necessary for the water to be stored in a tank above the collector

Computing

Thermosiphons are used for watercooling internal computer components, most commonly the processor. While any suitable liquid can be used, water is the easiest liquid to use in thermosiphon systems. Unlike traditional watercooling systems, thermosiphon systems do not rely on a pump but on convection for the movement of heated water (which may become vapour) from the components upwards to a heat exchanger. There the water is cooled and is ready to be recirculated. The most commonly used heat exchanger is a radiator, where air is blown actively through a fan system to condense the vapour to a liquid. The liquid is recirculated through the system, thus repeating the process. No pump is required—the vaporization and condensation cycle is self-sustaining.

Uses

Without cooling, modern processors can get hot to the point where they malfunction. Even with a common heat sink and fan cooling the processor, operating temperatures may still reach up to 70 °C (160 °F). A thermosiphon can handle heat output at a much wider temperature range than any heat sink and fan, and can maintain the processor 10–20 °C cooler. In some cases a thermosiphon may also be less bulky than a normal heat sink and fan.

Drawbacks

Thermosiphons must be mounted such that vapor rises up and liquid flows down to the boiler, with no bends in the tubing for liquid to pool. Also, the thermosiphon's fan that cools the gas needs cool air to operate. The system has to be completely airtight; if not, the process of thermosiphon will not take effect and cause the water to only evaporate over a small period of time.

Engine Cooling

Early cars and motor vehicles used thermosiphon circulation to move cooling water between their cylinder block and radiator. As engine power increased, increased flow was required and so engine-driven pumps were added to assist circulation. More compact engines then used smaller radiators and required more convoluted flow patterns, so the circulation became entirely dependent on the pump and might even be reversed against the natural circulation. An engine cooled only by thermosiphon is also very sensitive to low coolant level, i.e., losing only a small amount of coolant stops the circulation; a pump driven system is much more robust and can typically handle a lower coolant level.

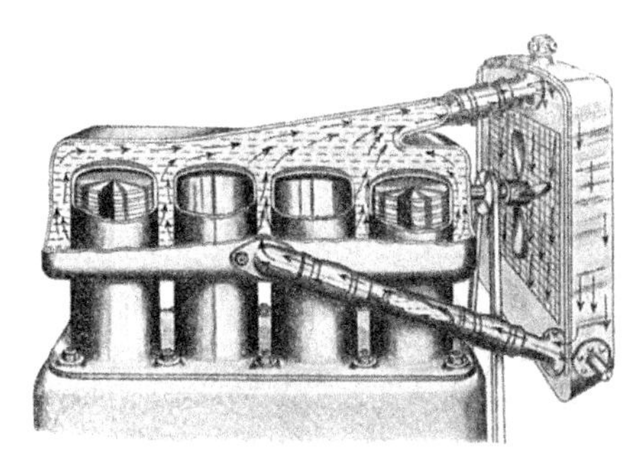

1937 diagram of engine cooling entirely by thermosiphon circulation

Thermal Reservoir

A thermal reservoir, a short-form of thermal energy reservoir, or thermal bath is a thermodynamic system with a heat capacity that is large enough that when it is in thermal contact with another system of interest or its environment, its temperature remains effectively constant. It is an effectively infinite pool of thermal energy at a given, constant temperature. The temperature of the reservoir does not change when heat is added or extracted because of the infinite heat capacity. As it can act as a source and sink of heat, it is often also referred to as a heat reservoir or heat bath.

Lakes, oceans and rivers often serve as thermal reservoirs in geophysical processes, such as the weather. In atmospheric science, large air masses in the atmosphere often function as thermal reservoirs.

The microcanonical partition sum $Z(E)$ of a heat bath of temperature T has the property

$$Z(E+\Delta E)=Z(E)e^{\Delta E/k_B T},$$

where k_B is Boltzmann's constant. It thus changes by the same factor when a given amount of energy is added. The exponential factor in this expression can be identified with the reciprocal of the Boltzmann factor.

References

- Perry, Robert H.; Green, Don W. (1984). Perry's Chemical Engineers' Handbook (6th ed.). McGraw-Hill. ISBN 0-07-049479-7.
- Sadik Kakaç; Hongtan Liu (2002). Heat Exchangers: Selection, Rating and Thermal Design (2nd ed.). CRC Press. ISBN 0-8493-0902-6.
- E.A.D.Saunders (1988). Heat Exchangers:Selection Design And Construction Longman Scientific and Technical ISBN 0-582-49491-5.

- Randall, David J.; Warren W. Burggren; Kathleen French; Roger Eckert (2002). Eckert animal physiology: mechanisms and adaptations. Macmillan. p. 587. ISBN 0-7167-3863-5.
- Kew, David Anthony Reay ; Peter. A. (2006). Heat pipes (5th ed.). Oxford: Butterworth-Heinemann. p. 309. ISBN 978-0-7506-6754-8.
- C, Yunus A.; Boles, Michael A. (2002). Thermodynamics: An Engineering Approach. Boston: McGraw-Hill. p. 247. ISBN 0-07-121688-X.
- Kew, David Anthony Reay ; Peter. A. (2006). Heat pipes (5th ed.). Oxford: Butterworth-Heinemann. p. 10. ISBN 978-0-7506-6754-8.
- Koen Crijns (2014-01-31). "Workshop Haswell delidding: improve CPU cooling!". hardware.info. Retrieved 2016-07-29.
- Haraburda, Scott S. (July 1995). "Three-Phase Flow? Consider Helical-Coil Heat Exchanger". Chemical Engineering. 102 (7): 149–151. Retrieved 14 July 2015.
- Rennie, Timothy J. (2004). Numerical And Experimental Studies Of A Doublepipe Helical Heat Exchanger (PDF) (Ph.D.). Montreal: McGill University. pp. 3–4. Retrieved 14 July 2015.
- Korane, Ashok B.; Purandare, P.S.; Mali, K.V. (June 2012). "Heat Transfer Analysis Of Helical Coil Heat Exchanger With Circular And Square Coiled Pattern" (PDF). International Journal of Engineering & Science Research. 2 (6): 413–423. Retrieved 14 July 2015.
- Kuvadiya, Manish N.; Deshmukh, Gopal K.; Patel, Rankit A.; Bhoi, Ramesh H. (April 2015). "Parametric Analysis of Tube in Tube Helical Coil Heat Exchanger at Constant Wall Temperature" (PDF). International Journal of Engineering Research & Technology. 1 (10): 279–285. Retrieved 14 July 2015.
- Patil, Ramachandra K.; Shende, B.W.; Ghosh, Prasanfa K. (13 December 1982). "Designing a helical-coil heat exchanger". Chemical Engineering. 92 (24): 85–88. Retrieved 14 July 2015.
- "Improving materials that convert heat to electricity and vice-versa". Ecnmag.com. May 6, 2013. Retrieved 2013-05-07.

Permissions

All chapters in this book are published with permission under the Creative Commons Attribution Share Alike License or equivalent. Every chapter published in this book has been scrutinized by our experts. Their significance has been extensively debated. The topics covered herein carry significant information for a comprehensive understanding. They may even be implemented as practical applications or may be referred to as a beginning point for further studies.

We would like to thank the editorial team for lending their expertise to make the book truly unique. They have played a crucial role in the development of this book. Without their invaluable contributions this book wouldn't have been possible. They have made vital efforts to compile up to date information on the varied aspects of this subject to make this book a valuable addition to the collection of many professionals and students.

This book was conceptualized with the vision of imparting up-to-date and integrated information in this field. To ensure the same, a matchless editorial board was set up. Every individual on the board went through rigorous rounds of assessment to prove their worth. After which they invested a large part of their time researching and compiling the most relevant data for our readers.

The editorial board has been involved in producing this book since its inception. They have spent rigorous hours researching and exploring the diverse topics which have resulted in the successful publishing of this book. They have passed on their knowledge of decades through this book. To expedite this challenging task, the publisher supported the team at every step. A small team of assistant editors was also appointed to further simplify the editing procedure and attain best results for the readers.

Apart from the editorial board, the designing team has also invested a significant amount of their time in understanding the subject and creating the most relevant covers. They scrutinized every image to scout for the most suitable representation of the subject and create an appropriate cover for the book.

The publishing team has been an ardent support to the editorial, designing and production team. Their endless efforts to recruit the best for this project, has resulted in the accomplishment of this book. They are a veteran in the field of academics and their pool of knowledge is as vast as their experience in printing. Their expertise and guidance has proved useful at every step. Their uncompromising quality standards have made this book an exceptional effort. Their encouragement from time to time has been an inspiration for everyone.

The publisher and the editorial board hope that this book will prove to be a valuable piece of knowledge for students, practitioners and scholars across the globe.

Index

CPSIA information can be obtained
at www.ICGtesting.com
Printed in the USA
BVOW07*1202061017
496951BV00005B/16/P